First-Principles Calculations
in Real-Space Formalism

First-Principles Calculations in Real-Space Formalism

Electronic Configurations and Transport Properties of Nanostructures

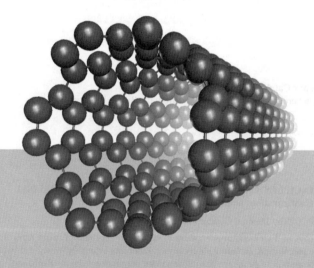

Kikuji Hirose
Osaka University, Japan

Tomoya Ono
Osaka University, Japan

Yoshitaka Fujimoto
University of Tokyo, Japan

Shigeru Tsukamoto
National Institute for Materials Science, Japan

ICP

Imperial College Press

phys
013713814

Published by

Imperial College Press
57 Shelton Street
Covent Garden
London WC2H 9HE

Distributed by

World Scientific Publishing Co. Pte. Ltd.
5 Toh Tuck Link, Singapore 596224
USA office: 27 Warren Street, Suite 401-402, Hackensack, NJ 07601
UK office: 57 Shelton Street, Covent Garden, London WC2H 9HE

British Library Cataloguing-in-Publication Data
A catalogue record for this book is available from the British Library.

FIRST-PRINCIPLES CALCULATIONS IN REAL-SPACE FORMALISM
Electronic Configurations and Transport Properties of Nanostructures

ISBN 1-86094-512-0

Printed in Singapore by World Scientific Printers (S) Pte Ltd

Preface

Currently, as various functional materials and minute electronic devices are produced by the latest nanoscale fabrication technology, it is quite indispensable for scientists and engineers, who continue to challenge the frontier of Nanotechnology, to study and/or rely on authentic first-principles (*ab initio*) calculation methods for correctly understanding electronic configurations and transport properties of nanostructures. However, we know a few practical and tractable calculation methods that accurately describe the relevant physics in nanostructures suspended between semi-infinite electrodes. Indeed, the plane-wave expansion method, a prevailing calculation method in solid-state physics, is not sufficient for determining current flow through nonperiodic nanostructures, because it requires artificial calculation models with three-dimensional periodic geometry. The tight-binding method using a basis set of atomic orbitals is applicable to the analysis for large systems; however, it fails in precise descriptions of electronic states, particularly in the region where tunneling effects dominate.

This book is based on our personal experience with electronic-transport calculations within the schemes of wave-function matching and Green's function matching. For some years, using real-space formalism free from any structural restrictions, we have been developing a first-principles calculation method for electronic-transport properties of nanostructures—overbridging boundary-matching (OBM) method, and we have found that the OBM method is a simple and practical method for doing highly accurate calculations. The book covers this method in a fairly complete fashion, besides introducing other notable real-space calculation methods.

The book consists of two parts: Part I (Chapters 1–5) contains the basic formalism of the real-space finite-difference method with its applications, which lays the vital theoretical foundations of the OBM method given in Part II. The real-space finite-difference method, in which wave functions and potentials are directly evaluated on real-space grid points instead of using any basis-function sets, has tackled the serious drawbacks of the plane-wave approach, e.g., its inability to describe strictly nonperiodic systems. The *flexibility* of this method allows for highly efficient solutions of electronic ground states of systems in great variety. Then, Part II (Chapters 6–10) focuses on the methods for calculating electronic-transport properties of nanostructures sandwiched between semi-infinite electrodes. Chapters 6–8 are devoted to the formulation of the OBM procedure for wave-function matching and its applications, whereas Chapter 9 deals with some fundamentals of the Green's function formalism in view of actual practical use. In Appendix B, one sees that when combined with the OBM procedure, the tight-binding method can be a powerful tool for examining transport properties of still larger systems. We attempt to be fairly complete as regards basic schemes, and try to be fairly rigorous with individual subjects.

The authors express their sincere thanks to their coworkers, Dr. M. Otani, Dr. Y. Egami, Dr. T. Sasaki, and many graduate students for their collaboration in constructing this framework and examining its potential in various applications. Several important pieces of material in this book have been produced since 1996 from the activities of the authors' group in the 'Center of Excellence' project (No. 08CE2004) of the Japanese Ministry of Education, Culture, Sports, Science and Technology. Finally, one of the authors (K. Hirose) is grateful to his wife Noriko for the typing of the manuscript and her continual support.

Osaka, September 2004

Kikuji Hirose

Contents

Real-Space Finite-Difference Method for First-Principles Calculations

PART 1

Real-Space Finite-Difference Method
for First-Principles Calculations

Chapter 1

Foundations of Methodology

Tomoya Ono and Kikuji Hirose

1.1 Real-Space Finite-Difference Method

Today, the performance of computers is highly advanced, and we have reached the conclusion that the use of first-principles calculation is the most appropriate for highly accurate analysis of problems at the atomic or electronic scale. For the current first-principles calculation, high-speed simulation with high accuracy is necessary. Although the conventional methods using basis set such as plane waves or atomic orbitals have achieved some success [see, for example, Car and Parrinello (1985), Szabo and Ostlund (1989), Payne *et al.* (1992), and Jensen (1999)], we should always pay attention to the fact that the boundary condition does not correspond to that of the actual experiments and to whether the basis function used satisfies the required calculation accuracy. On the other hand, with the real-space finite-difference method, since the wave function and potential on real-space grids are directly calculated instead of using basis functions, the above problems are avoided. Furthermore, because it has the following advantages compared with conventional methods, the real-space finite-difference method has recently become the focus of attention.

(i) Since arbitrary boundary conditions are available, the method can treat a model that corresponds to the actual experiment. In addition, simulation in which an electric and/or magnetic field is applied is possible.

(ii) In order to improve the calculation accuracy, the grid spacing should be narrowed, the procedure for which is simple and definite.

(iii) Since most of the values of the Hamiltonian matrix elements are

3

zero, one can easily create a calculation program.

(iv) Since all of the calculations are carried out in 'real space', it is easy to incorporate a localized orbital, which is localized in a finite region, required for so-called order-N $[O(N)]$ calculation, in which the amount of calculations is proportional to the model size. See, for example, Galli and Parrinello (1992), Aoki (1993), Li *et al.* (1993), Ordejón *et al.* (1993, 1995a, 1995b, 1996), Galli and Mauri (1994), Hierse and Stechel (1994), Nunes and Vanderbilt (1994), Stechel *et al.* (1994), Wang and Zunger (1994), Carlsson (1995), Faulkner *et al.* (1995), Hernández and Gillan (1995), Itoh *et al.* (1995, 1996), Wang *et al.* (1995), Hernández *et al.* (1996), Baer and Head-Gordon (1997), Millam and Scuseria (1997), Yang (1997), Bates *et al.* (1998), Stephan *et al.* (1998), York *et al.* (1998), Abdurixit (1999), Challacombe (1999), Roche (1999), Galli (2000), Hoshi and Fujiwara (2000, 2003), Ozaki *et al.* (2000), Ozaki (2001), Ozaki and Terakura (2001), and Sasaki *et al.* (2004).

(v) Since the method does not use fast Fourier transforms, parallel algorithms are easily achieved, enabling high-speed calculation using a massively parallel computer.

(vi) It is easy to develop the formalism used to solve the time-dependent Schrödinger equation. See, for example, Nomura *et al.* (1996, 1998, 1999), Sugino and Miyamoto (1999), Bertsch *et al.* (2000), and Watanabe and Tsukada (2000a, 2000b).

In the following sections, fundamental knowledge on the real-space finite-difference method is explained on the basis of the scheme of the density-functional theory. For a review, see Beck (2000). Atomic units $|e| = m = h/2\pi = 1$ are employed throughout this book (unless otherwise specified), where e, m and h are the electron charge, electron mass and Planck's constant, respectively.

1.2   Density-Functional Theory and Kohn–Sham Equation

We briefly introduce the central equations of the density-functional theory for electronic structure calculations. The most difficult problem in any electronic structure calculation is posed by the need to take account of the effects of electron-electron interaction. Hohenberg and Kohn (1964) provided the underlying theorems showing that for the ground state of the many-electron system, the Thomas–Fermi model [Thomas (1927); Fermi (1927, 1928a, 1928b); March (1975)] can be viewed as an approximation to an exact theory, the density-functional theory. In the theory, the ground-state energy of the system is the minimum value of the total-energy functional, and the density that yields this minimum value completely determines *all* properties of the system, such as all eigenfunctions and eigenvalues. Later, Kohn and Sham (1965) invented an ingenious approach to obtain the ground-state density by solving a set of accessible one-electron self-consistent eigenvalue equations. For more details on the density-functional theory, see von Barth (1984), Dreizler and da Providencia (1985), Jones and Gunnarsson (1989), Parr and Yang (1989), and Kryachko and Ludena (1990).

The Kohn–Sham total-energy functional is given as

$$E_{tot} = 2 \sum_i^M n_i \int_\Omega \psi_i^*(r) \left(-\frac{1}{2}\nabla^2 \right) \psi_i(r) dr$$

$$+ 2 \sum_s \sum_i^M n_i \int_\Omega \psi_i^*(r) v_{ion}^s(r - R^s)\psi_i(r) dr + \int_\Omega v_f(r)\rho(r) dr$$

$$+ \frac{1}{2} \int_\Omega \int \frac{\rho(r)\rho(r')}{|r - r'|} dr' dr + \int_\Omega \varepsilon_{xc}(r)\rho(r) dr - \sum_s Z_s v_f(R^s) + \gamma_E,$$

$$(1.1)$$

where v_{ion}^s is the ionic pseudopotential describing the interaction between the s-th ion located at position R^s and electron, v_f is the external-electric-field potential, ε_{xc} is the exchange-correlation potential, γ_E is the Coulomb energy associated with interactions among the nuclei (see Appendix A), Z_s is the valence of the s-th atom, M is the number of total wave functions, and $\rho(r)$ is the electron density distribution,

$$\rho(r) = 2 \sum_i^M n_i \int_\Omega |\psi_i(r)|^2 dr. \qquad (1.2)$$

Here, $\int_\Omega f(r)dr$ represents the integration of a function $f(r)$ inside the calculation domain Ω. In addition, n_i is the occupation number of each wave function and includes a k-point weight if there is more than one k-point for the integration in the Brillouin zone. Although, for simplicity, we here assume that the contributions from up and down spin electrons are the same, the inclusion of the effect of spin polarization is straightforward.

It is necessary to determine the set of wave functions ψ_i that minimize the Kohn–Sham total-energy functional, which is given by the self-consistent solution of the Kohn–Sham equations,

$$\left[-\frac{1}{2}\nabla^2 + v_{eff}(r) \right] \psi_i(r) = \varepsilon_i \psi_i(r), \qquad (1.3)$$

where

$$v_{eff}(\boldsymbol{r}) = \sum_s v_{ion}^s(\boldsymbol{r} - \boldsymbol{R}^s) + v_f(\boldsymbol{r}) + \int \frac{\rho(\boldsymbol{r}')}{|\boldsymbol{r} - \boldsymbol{r}'|} d\boldsymbol{r}' + v_{xc}(\boldsymbol{r}). \qquad (1.4)$$

Here, the first, second, third, and fourth terms are the ionic pseudopotential, external-electric-field potential, Hartree potential, and exchange-correlation potential, respectively.

1.3 Finite-Difference Formulas

Differentiation of wave functions and electron density distribution is one of the frequently performed processes in first-principles molecular-dynamics simulations. In the conventional methods using basis functions, the derivative can be obtained by differentiating basis functions. However, the prime concept of the real-space finite-difference method does not include basis functions; thus, the k-th order derivative of a function $f(x)$ at a grid point $x = ih_x$ (i: integer) is approximated by the following finite-difference formulas [Chelikowsky *et al.* (1994a, 1994b, 1996); Jing *et al.* (1994)].

$$\left. \frac{d^{(k)}}{dx^{(k)}} f(x) \right|_{x=ih_x} \approx \sum_{n=-N_f}^{N_f} c_n^{(k)} f(ih_x + nh_x), \qquad (1.5)$$

where N_f and h_x represent the parameters determining the order of the finite-difference approximation and the grid spacing, respectively. The larger the N_f, the more accurate the level of finite-difference approximation. Since a computational effort increases in proportion to N_f, a value of $1 - 4$ is usually used. The weight in (1.5), $c_n^{(k)}$, is determined using the Taylor expansion. For example, when $N_f = 1$, i.e., in the case of the central finite difference, by expanding $f(ih_x - h_x)$ and $f(ih_x + h_x)$ around ih_x as,

$$f(ih_x - h_x) \approx f(ih_x) - \frac{f'(ih_x)}{1!} h_x + \frac{f''(ih_x)}{2!} h_x^2 + O(h_x^3), \qquad (1.6)$$

$$f(ih_x + h_x) \approx f(ih_x) + \frac{f'(ih_x)}{1!} h_x + \frac{f''(ih_x)}{2!} h_x^2 + O(h_x^3), \qquad (1.7)$$

and by subtracting (1.6) from (1.7), the first-order derivative

$$f'(ih_x) \approx \frac{-f(ih_x - h_x) + f(ih_x + h_x)}{2h_x}, \qquad (1.8)$$

and by adding (1.6) to (1.7), the second-order derivative

$$f''(ih_x) \approx \frac{f(ih_x - h_x) - 2f(ih_x) + f(ih_x + h_x)}{h_x^2} \tag{1.9}$$

are obtained. Hence, from the above equations, in the case of $N_f = 1$, the coefficients for the first-order derivative are found to be $c_{-1}^{(1)} = -1/2h_x$, $c_0^{(1)} = 0$ and $c_1^{(1)} = 1/2h_x$. Those for the second-order derivative are $c_{-1}^{(2)} = c_1^{(2)} = 1/h_x^2$ and $c_0^{(2)} = -2/h_x^2$. Refer to Chelikowsky *et al.* (1994b) for the values of coefficients $c_n^{(k)}$ when $N_f > 1$. For simplicity, hereafter, the case for $N_f = 1$, i.e., the central finite difference, is presented.

1.4 Real-Space Representation of Kohn–Sham Equation

In the real-space finite-difference method, the values of wave functions and the electron density distribution are given only on discrete grid points in real space. Accordingly, the Kohn–Sham Hamiltonian [Kohn and Sham (1965)] acting on a wave function must be given in a form discretized in real space. In this section, the procedure for the discretization of the Kohn–Sham Hamiltonian, according to the finite-difference approximation discussed in the preceding section, is demonstrated. For an example, the case of a one-dimensional model is described.

First, we consider the discretization of the kinetic-energy operator, i.e., the expansion of the Laplacian. When the calculation domain is uniformly divided into N grids and a *nonperiodic* (an isolated) boundary condition is imposed, the kinetic-energy operator is written as

$$-\frac{1}{2}\nabla^2\psi \approx -\frac{1}{2} \begin{pmatrix} c_0 & c_1 & 0 & 0 & \cdots & 0 & 0 \\ c_1 & c_0 & c_1 & 0 & & & 0 \\ 0 & \ddots & \ddots & \ddots & \ddots & & \vdots \\ \vdots & \ddots & c_1 & c_0 & c_1 & \ddots & \vdots \\ \vdots & & \ddots & \ddots & \ddots & \ddots & 0 \\ 0 & & & 0 & c_1 & c_0 & c_1 \\ 0 & 0 & \cdots & 0 & 0 & c_1 & c_0 \end{pmatrix} \begin{pmatrix} \psi(h_x) \\ \psi(2h_x) \\ \vdots \\ \psi(ih_x) \\ \vdots \\ \psi(Nh_x - h_x) \\ \psi(Nh_x) \end{pmatrix}. \tag{1.10}$$

Here and hereafter, $c_n = c_n^{(2)}$.

In the case of a *periodic* boundary condition,

$$
-\frac{1}{2}\nabla^2\psi \approx -\frac{1}{2}
\begin{pmatrix}
c_0 & c_1 & 0 & 0 & \cdots & 0 & c_1 e^{-ik_x L_x} \\
c_1 & c_0 & c_1 & 0 & & & 0 \\
0 & \ddots & \ddots & \ddots & \ddots & & \vdots \\
\vdots & & \ddots & c_1 & c_0 & c_1 & \ddots & \vdots \\
\vdots & & & \ddots & \ddots & \ddots & \ddots & 0 \\
0 & & & & 0 & c_1 & c_0 & c_1 \\
c_1 e^{ik_x L_x} & 0 & \cdots & 0 & 0 & c_1 & c_0
\end{pmatrix}
\begin{pmatrix}
\psi(h_x) \\
\psi(2h_x) \\
\vdots \\
\psi(ih_x) \\
\vdots \\
\psi(Nh_x - h_x) \\
\psi(Nh_x)
\end{pmatrix},
$$

$$(1.11)$$

where k_x and L_x represent the Bloch wave number and the length of the calculation domain (supercell), respectively.

Next, we discuss the inner product of the potential term in the Kohn–Sham Hamiltonian and the wave function. The potential term consists of the following four terms: ionic pseudopotential v_{ion}^s, external electric field potential v_f, Hartree potential v_H, and exchange-correlation potential v_{xc}. The ionic pseudopotential is divided into local and nonlocal components when the norm-conserving pseudopotential given by Bachelet *et al.* [Hamann *et al.* (1979); Bachelet *et al.* (1982)], or Troullier and Martins (1991) is employed. The norm-conserving pseudopotentials are explained in Section 1.5. When the separable form given by Kleinman and Bylander (1982) is used as a nonlocal component, the inner product between the pseudopotential and wave function is given by

$$
v_{ion}^s(ih_x - R_x^s)\psi(ih_x) = v_{loc}^s(ih_x - R_x^s)\psi(ih_x)
$$
$$
+ \sum_{lm} G_{lm}^s \hat{v}_l^s(ih_x - R_x^s)\psi_{lm}^{ps,s}(ih_x - R_x^s).
$$

$$(1.12)$$

Here,

$$
G_{lm}^s = \frac{\sum_{i=1}^N \psi_{lm}^{ps,s*}(ih_x - R_x^s)\hat{v}_l^s(ih_x - R_x^s)\psi(ih_x)h_x}{\langle \psi_{lm}^{ps,s}|\hat{v}_l^s|\psi_{lm}^{ps,s}\rangle}, \qquad (1.13)
$$

and R_x^s is the position of the nucleus. The first term and the second term on the right-hand side of (1.12) are called the local component and nonlocal component, respectively. Furthermore, l and m are orbital and azimuthal angular-momentum quantum numbers, respectively, and $\psi_{lm}^{ps,s}$ is the pseudo-wave function used to generate the pseudopotential.

The external electric-field potential which exerts on an electron located at position x is represented by the external electric field $E_f(x)$, i.e.,

$$v_f(x) = \int_{-\infty}^{x} E_f(x')dx'. \tag{1.14}$$

Eventually, the one-dimensional Kohn–Sham equation discretized in real space is given by

$$-\frac{1}{2} \sum_{n=-N_f}^{N_f} c_n \psi(ih_x + nh_x)$$

$$+ \left(\sum_s v_{ion}^s (ih_x - R_x^s) + v_f(ih_x) + v_H(ih_x) + v_{xc}(ih_x) \right) \psi(ih_x) = \varepsilon \psi(ih_x). \tag{1.15}$$

The above argument is straightfowardly extended to the case of the three-dimensional Kohn–Sham equation as

$$-\frac{1}{2} \sum_{n=-N_f}^{N_f} [c_{x,n} \psi(ih_x + nh_x, jh_y, kh_z) + c_{y,n} \psi(ih_x, jh_y + nh_y, kh_z)$$

$$+ c_{z,n} \psi(ih_x, jh_y, kh_z + nh_z)] + v_{eff}(ih_x, jh_y, kh_z) \psi(ih_x, jh_y, kh_z)$$

$$= \varepsilon \psi(ih_x, jh_y, kh_z), \tag{1.16}$$

where

$$v_{eff}(ih_x, jh_y, kh_z) = \sum_s v_{ion}^s (ih_x - R_x^s, jh_y - R_y^s, kh_z - R_z^s)$$

$$+ v_f(ih_x, jh_y, kh_z) + v_H(ih_x, jh_y, kh_z)$$

$$+ v_{xc}(ih_x, jh_y, kh_z). \tag{1.17}$$

1.5 Norm-Conserving Pseudopotentials

The calculations including inner-shell electrons demand a significantly high cutoff energy, i.e., a small grid spacing is required; this prevents us from implementing practical simulations. In addition, in the algorithm used in first-principles calculations, the computational cost is proportional to the square of the number of electrons involved. Thus, it is disadvantageous to include a large number of electrons. Furthermore, in reality, the behavior of inner-shell electrons in molecules and crystals is similar to that of electrons in isolated atoms; valence electrons in atoms constituting a material determine most of the properties of the material. On the basis of this observation, the pseudopotential method which treats valence electrons alone is commonly used in first-principles calculations [Phillips (1958); Cohen and Heine (1970); Joannopoulos *et al.* (1977); Redondo *et al.* (1977); Starkloff and Joannopoulos (1977); Zunger and Cohen (1979); Hamann *et al.* (1979); Kerker (1980); Bachelet *et al.* (1982); Yin and Cohen (1982); Hamann (1989); Shirley *et al.* (1989); Vanderbilt (1990); Troullier and Martins (1991); Wang and Zunger (1995); Goedecker *et al.* (1996); Wang and Stott (2003)].

The most frequently used pseudopotential in first-principles calculations is the norm-conserving pseudopotential technique developed by Bachelet *et al.* [Hamann *et al.* (1979); Bachelet *et al.* (1982)], as well as the improved version by Troullier and Martins (1991). The pseudo-wave function obtained using these pseudopotentials has the following characteristics (see Fig. 1.1).

    a. There are no nodes in the pseudo-wave function.

    b. The pseudo-wave function agrees with the all-electron wave function outside the inner-shell radius of r_c.

    c. The eigenvalue of the valence electron state using pseudopotentials is in accordance with the eigenvalue calculated including the inner-shell electrons.

    d. The norm $\int_0^{r_c} |u_l^{ps}(r)|^2 dr$ up to the inner-shell radius of r_c of the pseudo-wave function coincides with the norm of the all-electron wave function up to r_c. Here, $u_l(r)$ is equal to $r\varphi_l(r)$, which is obtained by multiplying the radial component $\varphi_l(r)$ of the wave function $\psi_{lm}(\boldsymbol{r}) = \varphi_l(r)Y_{lm}(\theta, \phi)$ by the distance from the nucleus r. Also, $Y_{lm}(\theta, \phi)$ is a spherical harmonic function.

The following is the outline of the preparation procedure for pseudopotentials.

    1. Obtain the one-electron state by solving the Kohn–Sham equation under a spherically symmetric field which includes the inner-shell electrons. Then, the all-electron wave function of valence electrons $u_{lm}^{ae}(r)$ and the potential $V^{ae}(r)$ are obtained.

    2. Produce the pseudo-wave function $u_l^{ps}(r)$ which satisfies the conditions **a.** – **d.** above for the valence electron having the angular momentum l.

    3. Generate pseudopotential $V_l^{ps}(r)$ by eliminating the Hartree and exchange-correlation potentials of valence electrons from $V^{ae}(r)$.

As shown below, the constructed pseudopotential acts on each angular mo-

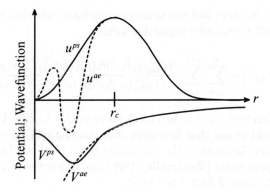

Fig. 1.1 Conceptual diagram of pseudopotential V^{ps} and pseudo-wave function u^{ps}. Here, V^{ae} is the potential obtained by the all-electron calculation; u^{ae} is the all-electron wave function.

mentum l as an operator.

$$V^{ps}(\boldsymbol{r}) = \sum_{l=0}^{\infty} \sum_{m=-l}^{+l} |Y_{lm}(\theta, \phi)\rangle V_l^{ps}(r) \langle Y_{lm}(\theta, \phi)|. \tag{1.18}$$

Usually, l is cut off at about 2. Furthermore, assuming an appropriate component $v_{loc}(r)$ as a local component and assuming the difference between each component $V_l^{ps}(r)$ and $v_{loc}(r)$,

$$\hat{v}_l(r) = V_l^{ps}(r) - v_{loc}(r), \tag{1.19}$$

to be the nonlocal component, the following form can be used:

$$V^{ps}(\boldsymbol{r}) = v_{loc}(r) + \sum_{l=0}^{\infty} \sum_{m=-l}^{+l} |Y_{lm}(\theta, \phi)\rangle \hat{v}_l(r) \langle Y_{lm}(\theta, \phi)|. \tag{1.20}$$

In particular, in the pseudopotential given by Bachelet *et al.* [Hamann *et al.* (1979); Bachelet *et al.* (1982)], the local component is given as

$$v_{loc}(\boldsymbol{r}) = -Z \left(\frac{C_1 \, \text{erf}(\sqrt{\alpha_1} \, |\boldsymbol{r}|)}{|\boldsymbol{r}|} + \frac{C_2 \, \text{erf}(\sqrt{\alpha_2} \, |\boldsymbol{r}|)}{|\boldsymbol{r}|} \right). \tag{1.21}$$

Here, $Z(> 0)$ is the sum of charges of the valence electrons, and $\text{erf}(x)$ is the error function (or probability integral) defined by

$$\text{erf}(x) = \frac{2}{\sqrt{\pi}} \int_0^x e^{-t^2} dt. \tag{1.22}$$

The form of (1.20) is called the nonseparable form, whereas Kleinman and Bylander (1982) proposed a separable form,

$$V^{ps}(\boldsymbol{r}) = v_{loc}(r) + \sum_{l=0}^{\infty} \sum_{m=-l}^{+l} \frac{|\psi_{lm}^{ps}(r,\theta,\phi)\hat{v}_l(r)\rangle\langle\hat{v}_l(r)\psi_{lm}^{ps}(r,\theta,\phi)|}{\langle\psi_{lm}^{ps}(r,\theta,\phi)|\hat{v}_l(r)|\psi_{lm}^{ps}(r,\theta,\phi)\rangle}. \quad (1.23)$$

The pseudopotential v_{ion} in (1.12) corresponds to this V^{ps}. It is not an overstatement to say that first-principles calculations have been realized owing to the development of the pseudopotential method. In recent years, softer pseudopotentials [Vanderbilt (1990)] in which the norm-conserving condition is eliminated have been used.

1.6 Hellmann–Feynman Forces Acting on Atoms

The implementation of first-principles molecular-dynamics simulations requires the Hellmann-Feynman forces acting on nuclei, which can be obtained by differentiating the total energy (1.1) with respect to the positions of the nuclei $\boldsymbol{R}^s$. Among the components in the total energy, the ionic pseudopotential energy, external electric-field potential energy of nuclei, and Coulomb energy γ_E are functions of the position of the nucleus.

First, the ionic pseudopotential energy E_{ne} is divided into two terms:

$$E_{ne} = E_{loc} + E_{nonlocal}, \quad (1.24)$$

and the first term due to the local components per supercell is represented by

$$E_{loc} = 2 \sum_s^M \sum_i n_i \int_\Omega \psi_i^*(\boldsymbol{r}) v_{loc}^s(\boldsymbol{r} - \boldsymbol{R}^s) \psi_i(\boldsymbol{r}) d\boldsymbol{r}. \quad (1.25)$$

Then, by changing the variable as $\boldsymbol{r} \to \boldsymbol{r}' = \boldsymbol{r} - \boldsymbol{R}^s$ and differentiating E_{loc} with respect to the position of the s-th nucleus $\boldsymbol{R}^s$, we have

$$-\frac{\partial E_{loc}}{\partial \boldsymbol{R}^s} = -2 \sum_i^M n_i \frac{\partial}{\partial \boldsymbol{R}^s} \int_\Omega \psi_i^*(\boldsymbol{r}' + \boldsymbol{R}^s) v_{loc}^s(\boldsymbol{r}') \psi_i(\boldsymbol{r}' + \boldsymbol{R}^s) d\boldsymbol{r}'$$

$$= -2 \sum_i^M n_i \left(\int_\Omega \frac{\partial \psi_i^*(\boldsymbol{r}' + \boldsymbol{R}^s)}{\partial \boldsymbol{R}^s} v_{loc}^s(\boldsymbol{r}') \psi_i(\boldsymbol{r}' + \boldsymbol{R}^s) d\boldsymbol{r}' \right.$$

$$\left. + \int_\Omega \psi_i^*(\boldsymbol{r}' + \boldsymbol{R}^s) v_{loc}^s(\boldsymbol{r}') \frac{\partial \psi_i(\boldsymbol{r}' + \boldsymbol{R}^s)}{\partial \boldsymbol{R}^s} d\boldsymbol{r}' \right). \quad (1.26)$$

Thus, with the reversion of the variable r' to the original form $r = r' + R^s$, (1.26) leads to

$$-\frac{\partial E_{loc}}{\partial R^s} = -2\sum_i^M n_i \left(\int_\Omega \frac{\partial \psi_i^*(r)}{\partial r} v_{loc}^s(r - R^s)\psi_i(r)dr \right.$$

$$\left. + \int_\Omega \psi_i^*(r')v_{loc}^s(r - R^s)\frac{\partial \psi_i(r)}{\partial r}dr \right)$$

$$= -4\sum_i^M n_i \, \text{Re} \int_\Omega \frac{\partial \psi_i^*(r)}{\partial r} v_{loc}^s(r - R^s)\psi_i(r)dr. \quad (1.27)$$

Next, using the separable form proposed by Kleinman and Bylander (1982), the second term due to the nonlocal components is given by

$$E_{nonloc} = 2\sum_s^M \sum_i \sum_{lm} \frac{W_{lm}^{s*}(R^s)W_{lm}^s(R^s)}{\langle \psi_{lm}^{ps,s}|\hat{v}_l^s|\psi_{lm}^{ps,s}\rangle}, \quad (1.28)$$

where

$$W_{lm}^s(R^s) = \int_\Omega \hat{v}_l^s(r - R^s)\psi_{lm}^{ps,s*}(r - R^s)\psi_i(r)dr, \quad (1.29)$$

for which, along the same line as the derivation of (1.27), we obtain

$$-\frac{\partial E_{nonloc}}{\partial R^s} = -2\sum_i^M n_i \sum_{lm} \frac{\partial}{\partial R^s} \frac{W_{lm}^{s*}(R^s)W_{lm}^s(R^s)}{\langle \psi_{lm}^{ps,s}|\hat{v}_l^s|\psi_{lm}^{ps,s}\rangle}$$

$$= -4\sum_i^M n_i \sum_{lm} \frac{1}{\langle \psi_{lm}^{ps,s}|\hat{v}_l^s|\psi_{lm}^{ps,s}\rangle}$$

$$\times \text{Re} \int_\Omega \hat{v}_l^s(r - R^s)\psi_{lm}^{ps,s}(r - R^s)\frac{\partial \psi_i^*(r)}{\partial r}dr W_{lm}^s(R^s),$$

$$(1.30)$$

where $\hat{v}_l^s(r)$ vanishes on the outside of the inner shell radius r_c of the pseudopotential, and the integration is performed only inside the inner shell radius.

In the real-space finite-difference formalism, the integrations are approximated by the summations on discrete grid points. An efficient technique for these summations is described in Chapter 4.

In addition, the derivative of the external electric field E_f of the nucleus is given by

$$-\frac{\partial[-Z_s v_f(\boldsymbol{R}^s)]}{\partial \boldsymbol{R}^s} = Z_s \boldsymbol{E}_f(\boldsymbol{R}^s). \tag{1.31}$$

Finally, although the derivative of the Coulomb energy among the nuclei is given by $-\partial \gamma_E/\partial \boldsymbol{R}^s$, the form of γ_E differs depending on the boundary conditions; the differentiation formulas are discussed in Appendix A.

Chapter 2

Solvers of the Poisson Equation and Related Techniques

Tomoya Ono and Kikuji Hirose

2.1 The Real-Space Representation of the Poisson Equation

The most direct approach for obtaining the Hartree potential is numerical integration using the following equation.

$$v_H(\boldsymbol{r}) = \int \frac{\rho(\boldsymbol{r}')}{|\boldsymbol{r} - \boldsymbol{r}'|} d\boldsymbol{r}'. \tag{2.1}$$

However, being proportional to the square of the model size, the direct integration is computationally very demanding. Therefore, in first-principles calculations, the Hartree potential is commonly determined by solving the Poisson equation:

$$\nabla^2 v_H(\boldsymbol{r}) = -4\pi\rho(\boldsymbol{r}). \tag{2.2}$$

In this section, a procedure for solving the Poisson equation using the real-space finite-difference method is explained. For simplicity, examples applying the central finite-difference formula, i.e., $N_f = 1$ in (1.5), to the one-dimensional case are presented.

The second-order derivative of the Hartree potential is approximated as,

$$\begin{aligned}
\nabla^2 v_H(ih_x) &\approx c_1 v_H(ih_x - h_x) + c_0 v_H(ih_x) + c_1 v_H(ih_x + h_x) \\
&= \frac{v_H(ih_x - h_x) - 2v_H(ih_x) + v_H(ih_x + h_x)}{h_x^2},
\end{aligned} \tag{2.3}$$

where h_x is the grid spacing. Here and hereafter, c_i is $c_i^{(2)}$ in (1.5). Thus, the Poisson equation (2.2) can be replaced by N-dimensional simultaneous equations, where N is the number of the grid points. When a nonperiodic (an isolated) boundary condition is imposed, the simultaneous equations are given by

$$-\frac{2v_H(h_x)}{h_x^2} + \frac{v_H(2h_x)}{h_x^2} = -4\pi\rho(h_x) - \frac{v_H(0)}{h_x^2},$$

$$\frac{v_H(h_x)}{h_x^2} - \frac{2v_H(2h_x)}{h_x^2} + \frac{v_H(3h_x)}{h_x^2} = -4\pi\rho(2h_x),$$

$$\vdots \qquad\qquad \vdots$$

$$\frac{v_H(ih_x-h_x)}{h_x^2} - \frac{2v_H(ih_x)}{h_x^2} + \frac{v_H(ih_x+h_x)}{h_x^2} = -4\pi\rho(ih_x),$$

$$\vdots \qquad\qquad \vdots$$

$$\frac{v_H(Nh_x-2h_x)}{h_x^2} - \frac{2v_H(Nh_x-h_x)}{h_x^2} + \frac{v_H(Nh_x)}{h_x^2} = -4\pi\rho(Nh_x - h_x),$$

$$\frac{v_H(Nh_x-h_x)}{h_x^2} - \frac{2v_H(Nh_x)}{h_x^2} = -4\pi\rho(Nh_x) - \frac{v_H(Nh_x+h_x)}{h_x^2},$$

$$(2.4)$$

and these are represented using a matrix as,

$$\begin{pmatrix} c_0 & c_1 & 0 & 0 & \cdots & 0 & 0 \\ c_1 & c_0 & c_1 & 0 & & & 0 \\ 0 & \ddots & \ddots & \ddots & \ddots & & \vdots \\ \vdots & \ddots & c_1 & c_0 & c_1 & \ddots & \vdots \\ \vdots & & \ddots & \ddots & \ddots & \ddots & 0 \\ 0 & & & 0 & c_1 & c_0 & c_1 \\ 0 & 0 & \cdots & 0 & 0 & c_1 & c_0 \end{pmatrix} \begin{pmatrix} \alpha_1 \\ \alpha_2 \\ \vdots \\ \alpha_i \\ \vdots \\ \alpha_{N-1} \\ \alpha_N \end{pmatrix} = \begin{pmatrix} \beta_1 \\ \beta_2 \\ \vdots \\ \beta_i \\ \vdots \\ \beta_{N-1} \\ \beta_N \end{pmatrix}, \qquad (2.5)$$

where $\alpha_i = v_H(ih_x)$ and

$$\beta_i = \begin{cases} -4\pi\rho(h_x) - \frac{1}{h_x^2}v_H(0) & i = 1 \\ -4\pi\rho(Nh_x) - \frac{1}{h_x^2}v_H(Nh_x + h_x) & i = N \\ -4\pi\rho(ih_x) & \text{otherwise.} \end{cases} \qquad (2.6)$$

Here $v_H(0)$ and $v_H(Nh_x + h_x)$ are the boundary values, i.e., the values of the Hartree potential just outside of the calculation domain. The method for determining the boundary values is explained in the following section.

Similarly, in the case of a periodic boundary condition, we have

$$\frac{v_H(Nh_x)}{h_x^2} - \frac{2v_H(h_x)}{h_x^2} + \frac{v_H(2h_x)}{h_x^2} = -4\pi\rho(h_x),$$

$$\frac{v_H(h_x)}{h_x^2} - \frac{2v_H(2h_x)}{h_x^2} + \frac{v_H(3h_x)}{h_x^2} = -4\pi\rho(2h_x),$$

$$\vdots \qquad\qquad \vdots$$

$$\frac{v_H(ih_x-h_x)}{h_x^2} - \frac{2v_H(ih_x)}{h_x^2} + \frac{v_H(ih_x+h_x)}{h_x^2} = -4\pi\rho(ih_x), \tag{2.7}$$

$$\vdots \qquad\qquad \vdots$$

$$\frac{v_H(Nh_x-2h_x)}{h_x^2} - \frac{2v_H(Nh_x-h_x)}{h_x^2} + \frac{v_H(Nh_x)}{h_x^2} = -4\pi\rho(Nh_x - h_x),$$

$$\frac{v_H(Nh_x-h_x)}{h_x^2} - \frac{2v_H(Nh_x)}{h_x^2} + \frac{v_H(h_x)}{h_x^2} = -4\pi\rho(Nh_x).$$

and by representing these using a matrix

$$\begin{pmatrix} c_0 & c_1 & 0 & 0 & \cdots & 0 & c_1 \\ c_1 & c_0 & c_1 & 0 & & & 0 \\ 0 & \ddots & \ddots & \ddots & \ddots & & \vdots \\ \vdots & \ddots & c_1 & c_0 & c_1 & \ddots & \vdots \\ \vdots & & \ddots & \ddots & \ddots & \ddots & 0 \\ 0 & & & 0 & c_1 & c_0 & c_1 \\ c_1 & 0 & \cdots & 0 & 0 & c_1 & c_0 \end{pmatrix} \begin{pmatrix} \alpha_1 \\ \alpha_2 \\ \vdots \\ \alpha_i \\ \vdots \\ \alpha_{N-1} \\ \alpha_N \end{pmatrix} = \begin{pmatrix} \beta_1 \\ \beta_2 \\ \vdots \\ \beta_i \\ \vdots \\ \beta_{N-1} \\ \beta_N \end{pmatrix}, \tag{2.8}$$

where $\alpha_i = v_H(ih_x)$ and $\beta_i = -4\pi\rho(ih_x)$. In the case of imposing periodic boundary condition, the boundary values are not required.

Furthermore, when higher-order finite-difference formulas, i.e., $N_f > 1$ in (1.5), are used, the relevant matrix elements A_{ij} and β for the non-periodic boundary condition are

$$A_{ij} = c_{|i-j|}, \tag{2.9}$$

and

$$\beta_i = \begin{cases} -4\pi\rho(ih_x) - \displaystyle\sum_{n=i}^{N_f} c_n v_H(ih_x - nh_x) & i \leq N_f \\[2ex] -4\pi\rho(ih_x) - \displaystyle\sum_{n=N-i+1}^{N_f} c_n v_H(ih_x + nh_x) & i \geq N - N_f + 1 \\[2ex] -4\pi\rho(ih_x) & \text{otherwise,} \end{cases} \tag{2.10}$$

respectively, whereas for the periodic boundary condition,

$$A_{ij} = \begin{cases} c_{N-|i-j|} & |i-j| \geq N - N_f \\ c_{|i-j|} & \text{otherwise,} \end{cases} \tag{2.11}$$

and

$$\beta_i = -4\pi\rho(ih_x). \tag{2.12}$$

The simultaneous equations of (2.5) and (2.8) can be solved by the steepest-descent or conjugate-gradient algorithm (see Section 2.5).

2.2 The Fuzzy Cell Decomposition and Multipole Expansion Technique

When a nonperiodic boundary condition is imposed, it is necessary to know in advance the boundary values of the Hartree potential just outside of the calculation domain for solving the Poisson equation. The most direct procedure for obtaining these boundary values is numerical integration using (2.1). However, the computational effort is proportioned to the 5/3 power of the model size; hence, this is not an efficient and practical method. Chelikowsky *et al.* (1994a, 1994b) proposed a procedure using a multipole expansion of the charge distribution around an arbitrary point r'':

$$\begin{aligned} v_H(r) &= \int \frac{\rho(r')}{|r - r'|} dr' \\ &= \sum_{l=0}^{\infty} \int \frac{\rho(r')}{|r - r''|} \left(\frac{|r' - r''|}{|r - r''|} \right)^l P_l(\cos\theta') dr' \\ &= \frac{\int \rho(r')dr'}{|r - r''|} + \sum_{\mu=x,y,z} p_\mu \cdot \frac{(r_\mu - r''_\mu)}{|r - r''|^3} \\ &\quad + \sum_{\mu,\nu=x,y,z} q_{\mu\nu} \cdot \frac{3(r_\mu - r''_\mu)(r_\nu - r''_\nu) - \delta_{\mu\nu}|r - r''|^2}{|r - r''|^5} + \cdots, \end{aligned} \tag{2.13}$$

where p_μ is dipole moment,

$$p_\mu = \int (r_\mu - r''_\mu)\rho(r)dr, \tag{2.14}$$

and

$$q_{\mu\nu} = \int \frac{1}{2}(r_\mu - r''_\mu)(r_\nu - r''_\nu)\rho(\boldsymbol{r})d\boldsymbol{r}. \tag{2.15}$$

Here, the functions $P_l(\cos\theta')$ $(l = 0, 1, 2, \dots)$ are the Legendre polynomials, and $\cos\theta'$ is expressed as

$$\cos\theta' = \frac{(\boldsymbol{r} - \boldsymbol{r}'') \cdot (\boldsymbol{r}' - \boldsymbol{r}'')}{|\boldsymbol{r} - \boldsymbol{r}''| \cdot |\boldsymbol{r}' - \boldsymbol{r}''|}. \tag{2.16}$$

Using this method, it is possible to make the computational cost proportional to the model size. However, the result largely depends on the position of $\boldsymbol{r}''$ because the expansion is carried out only around *one* arbitrary point. Here, we introduce the fuzzy cell decomposition and multipole expansion (FCD-MPE) method, which circumvents these problems.

First, weighting functions $\omega_s(\boldsymbol{r})$ for the multiple-center system centered at the s-th nucleus are prepared. These functions are the defining functions of Voronoi polyhedra Ω_s which provide the Wigner–Seitz cells and should be set to satisfy the following equations:

$$\sum_s \omega_s(\boldsymbol{r}) = 1, \tag{2.17}$$

and

$$\omega_s(\boldsymbol{r}) = \begin{cases} 1 & \in \Omega_s \\ 0 & \text{otherwise.} \end{cases} \tag{2.18}$$

The electronic charge distribution is divided into the charge distributions existing around the s-th nucleus using these weighting functions, $\omega_s(\boldsymbol{r})$.

$$\rho_s(\boldsymbol{r}) = \rho(\boldsymbol{r})\omega_s(\boldsymbol{r}), \tag{2.19}$$

and

$$\rho(\boldsymbol{r}) = \sum_s \rho_s(\boldsymbol{r}). \tag{2.20}$$

By performing the multipole expansion for each $\rho_s(\boldsymbol{r})$ centering around the position of each nucleus $\boldsymbol{R}^s$,

$$v_H(\boldsymbol{r}) = \sum_s \left(\frac{\int \rho_s(\boldsymbol{r}')d\boldsymbol{r}'}{|\boldsymbol{r} - \boldsymbol{R}^s|} + \sum_{\mu=x,y,z} p_{s,\mu} \cdot \frac{r_\mu - R^s_\mu}{|\boldsymbol{r} - \boldsymbol{R}^s|^3} \right.$$
$$\left. + \sum_{\mu,\nu=x,y,z} q_{s,\mu\nu} \cdot \frac{3(r_\mu - R^s_\mu)(r_\nu - R^s_\nu) - \delta_{\mu\nu}|\boldsymbol{r} - \boldsymbol{R}^s|^2}{|\boldsymbol{r} - \boldsymbol{R}^s|^5} + \cdots \right). \tag{2.21}$$

When $\omega_s(\boldsymbol{r})$ behaves like a step function at the boundary of Ω_s, many terms are required for the expansion of (2.21). In the FCD-MPE method, to carry out the expansion using as small a number of terms as possible, the behavior of $\omega_s(\boldsymbol{r})$ near the boundary is made smooth using the fuzzy cell technique [Becke (1988)], in which the section of the boundary of (2.18) is *fuzzy*. The fuzzy cell technique is explained in the next section. When the fuzzy cell is employed, the multipole expansion to $l = 2$ in (2.13) is sufficient to obtain an accurate boundary value.

2.3   Algorithm to Generate the Fuzzy Cell

We introduce an efficient algorithm for the automatic construction of Voronoi polyhedra. This procedure involves the popular two-center coordinate system known as conforcal elliptical coordinates (λ, μ, ϕ). With center i as reference, consider in turn each of the other centers $i \neq j$ and establish elliptical coordinates λ_{ij}, μ_{ij}, and ϕ_{ij} on the foci i and j. Of special interest is the coordinate

$$\mu_{ij} = \frac{r_i - r_j}{R_{ij}}, \tag{2.22}$$

where r_i and r_j denote distances to i-th and j-th nuclei, respectively, and R_{ij} is the internucleus distance. With this hyperboloidal coordinate as argument, consider the step function

$$s(\mu_{ij}) = \begin{cases} 1 & -1 \le \mu_{ij} \le 0 \\ 0 & 0 < \mu_{ij} \le 1. \end{cases} \tag{2.23}$$

Since the surface $\mu_{ij} = 0$ is the perpendicular bisector of R_{ij}, the Voronoi polyhedron on the i-th nucleus is defined by the following simple product:

$$P_i(\boldsymbol{r}) = \prod_{j \neq i} s(\mu_{ij}). \tag{2.24}$$

Let us replace the step function (2.23) with continuous smooth functions. We shall impose $s(\mu_{ij})$ the following minimum requirements:

$$s(-1) = 1,$$
$$s(1) = 0, \tag{2.25}$$

and

$$\frac{ds}{d\mu}\bigg|_{\mu=-1} = \frac{ds}{d\mu}\bigg|_{\mu=1} = 0. \tag{2.26}$$

The derivative constrains of (2.26) ensure that $s(\mu_{ij})$ does not have cusps. The simplest possible $s(\mu_{ij})$ satisfying the above constrains is obtained by polynomials:

$$f(\mu) = \frac{3}{2}\mu - \frac{1}{2}\mu^3,$$

$$s(\mu) = \frac{1}{2}[1 - f(\mu)]. \qquad (2.27)$$

The simple polynomial varies smoothly between the end points -1 and 1, but is not sharp. If we iterate as follows:

$$f_0(\mu) = f(\mu),$$
$$f_1(\mu) = f(f_0(\mu)),$$
$$\vdots$$
$$f_k(\mu) = f(f_{k-1}(\mu)), \qquad (2.28)$$

and

$$s_k(\mu) = \frac{1}{2}[1 - f_k(\mu)], \qquad (2.29)$$

then successively sharper functions may be generated with increasing iteration order k. According to Becke (1988), the value $k = 3$ is appropriate for general applications.

Finally, in order to satisfy the requirement of (2.17), we use the following definition:

$$\omega_s(\boldsymbol{r}) = \frac{P_s(\boldsymbol{r})}{\sum_{s'} P_{s'}(\boldsymbol{r})}, \qquad (2.30)$$

where the summation over s' in the denominator includes all nuclei in the system.

2.4 Illustration for Efficiency of the Fuzzy Cell Decomposition and Multipole Expansion Method

In order to demonstrate the accuracy and efficiency of the FCD-MPE method, the Poisson equation is solved using the boundary values obtained by the FCD-MPE method and the multipole expansion around *one* arbitrary point in the calculation domain. Here, the charge distribution is assumed to be

$$\rho(\boldsymbol{r}) = \sum_{i=1,2} Z_i \times \left(\frac{b_i}{\pi}\right)^{\frac{3}{2}} \exp[-b_i|\boldsymbol{r} - \boldsymbol{a}_i|^2], \qquad (2.31)$$

which imitates the charge distribution of a diatomic molecule, where $\boldsymbol{a}_1 = (2,0,0)$, $\boldsymbol{a}_2 = (-2,0,0)$, $b_1 = 0.8$, $b_2 = 0.6$, $Z_1 = 6$ and $Z_2 = 4$. The cell size is set at $18.0 \times 16.0 \times 16.0$ a.u., a nine-point finite difference formula, i.e., $N_f = 4$ in (1.5), is employed for the second-order derivative, and the grid spacing is chosen to be 0.15 a.u. In order to generate the Voronoi polyhedron within the framework of the fuzzy cell technique, the value of $k = 3$ is adopted in (2.28) and (2.29). Figure 2.1(A) shows the relative error for the cross section at $z = 0.075$ a.u. between the analytical solution, i.e.,

$$v_H(\boldsymbol{r}) = \sum_{i=1,2} Z_i \times \frac{\text{erf}\left(\sqrt{b_i}|\boldsymbol{r} - \boldsymbol{a}_i|\right)}{|\boldsymbol{r} - \boldsymbol{a}_i|}, \qquad (2.32)$$

and the computed Hartree potential using the boundary values obtained by FCD-MPE method. Here, $\text{erf}(x)$ is the error function (or probability integral) defined by

$$\text{erf}(x) = \frac{2}{\sqrt{\pi}} \int_0^x e^{-t^2} dt. \qquad (2.33)$$

We also depict in Fig. 2.1(B) the result using the multipole expansion around one point:

$$\boldsymbol{r}'' = \frac{\int \boldsymbol{r}'\rho(\boldsymbol{r}')d\boldsymbol{r}'}{\int \rho(\boldsymbol{r}')d\boldsymbol{r}'}. \qquad (2.34)$$

As seen in Fig. 2.1, these results make it clear that the capability of solving the Poisson equation is drastically improved by incorporating the FCD-MPE method into the real-space finite-difference scheme.

2.5   Conjugate-Gradient Method

We now describe the conjugate-gradient (CG) method devised by Hestenes and Stiefel (1952) as applied to the linear system $A\boldsymbol{x} = \boldsymbol{b}$ given by (2.5) and (2.8). Consider the action functional

$$F(\boldsymbol{x}) = \frac{1}{2}(\boldsymbol{x}, A\boldsymbol{x}) - (\boldsymbol{x}, \boldsymbol{b}), \qquad (2.35)$$

where A is a symmetric and positive definite $N_{xyz}(= N_x \times N_y \times N_z)$-dimensional matrix, and N_x, N_y and N_z are the numbers of grid points in

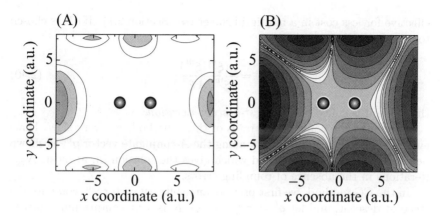

Fig. 2.1 Relative errors between the Hartree potential using boundary values computed by each multipole expansion method and that evaluated by the analytical solution for the cross section at $z = 0.075$ a.u. (A) Results using boundary values obtained by the FCD-MPE method. (B) Results using boundary values obtained by the multipole expansion around one point. Spheres show the locations of the nuclei. The maps are illustrated on logarithmic scale and each contour represents twice or half density of the adjacent contour lines. The lowest contour represents 0.0025 %.

the x, y and z directions, respectively. The problem of solving $A\boldsymbol{x} = \boldsymbol{b}$ is equivalent to the problem of minimizing $F(\boldsymbol{x})$, since

$$F(\boldsymbol{x}) = F(\hat{\boldsymbol{x}}) + \frac{1}{2}((\boldsymbol{x} - \hat{\boldsymbol{x}}), A(\boldsymbol{x} - \hat{\boldsymbol{x}})), \qquad (2.36)$$

where $\hat{\boldsymbol{x}}$ is the exact solution of $A\boldsymbol{x} = \boldsymbol{b}$. Taking the functional derivative of the action with respect to $\boldsymbol{x}$, the gradient of $F(\boldsymbol{x})$ is given by

$$\text{grad } F(\boldsymbol{x}) = A\boldsymbol{x} - \boldsymbol{b}. \qquad (2.37)$$

The direction of the vector, $-\text{grad } F(\boldsymbol{x})$, is the steepest-descent direction for which the functional $F(\boldsymbol{x})$ at point $\boldsymbol{x}$ has the greatest instantaneous rate of change. Let us look at the corresponding nonstationary process:

$$\boldsymbol{x}^{m+1} = \boldsymbol{x}^m + \alpha^m \boldsymbol{\xi}^m, \qquad (2.38)$$

where the superscript m indicates the iteration number and

$$\boldsymbol{\xi}^m = -\text{grad } F(\boldsymbol{x}^m) = \boldsymbol{b} - A\boldsymbol{x}^m. \qquad (2.39)$$

If α^m is arbitrarily chosen to satisfy the required stability criterion, the method is called the *Jacobi iteration*, and if the elements of the vector $\boldsymbol{x}$ are successively displaced, the method is called the *Gauss–Seidel iteration*. These simple relaxation procedures (particularly the Gauss–Seidel one) are effective for less cost in a multigrid solver (see Section 2.7). If α^m is chosen to minimize $F(\boldsymbol{x}^{m+1})$, i.e.,

$$\alpha^m = \frac{(\boldsymbol{\xi}^m, \boldsymbol{\xi}^m)}{(\boldsymbol{\xi}^m, A\,\boldsymbol{\xi}^m)}, \tag{2.40}$$

the method is called the *steepest-descent iteration*.

Generally, the convergence rates of these methods are very slow. However, by updating the vector $\boldsymbol{x}^m$ along the A-conjugate vector $\boldsymbol{g}^m$ as shown below, we obtain a CG method which gives the solution in, at most, N_{xyz} iterations in the absence of rounding errors.

In the CG method, we first prepare an arbitrary vector $\boldsymbol{x}^0$, compute the steepest descent, and let $\boldsymbol{g}^0 = K\boldsymbol{\xi}^0$, where K is a preconditioning matrix. The preconditioning technique of accelerating convergence will be explained in Section 2.6. K is a unit matrix in the case without preconditioning. Then we implement iteratively the following procedure from (2.41) to (2.45):

$$\boldsymbol{x}^{m+1} = \boldsymbol{x}^m + \alpha^m \boldsymbol{g}^m, \tag{2.41}$$

where α^m is chosen to minimize $F(\boldsymbol{x}^{m+1})$:

$$\alpha^m = \frac{(\boldsymbol{\xi}^m, K\boldsymbol{\xi}^m)}{(\boldsymbol{g}^m, A\boldsymbol{g}^m)}, \tag{2.42}$$

$$\boldsymbol{\xi}^{m+1} = -\mathrm{grad}F(\boldsymbol{x}^{m+1}) = \boldsymbol{b} - A\boldsymbol{x}^{m+1}, \tag{2.43}$$

and

$$\boldsymbol{g}^{m+1} = K\boldsymbol{\xi}^{m+1} + \beta^m \boldsymbol{g}^m, \tag{2.44}$$

for $m = 0, 1, 2, \ldots$. Here β^m is chosen so that $\boldsymbol{g}^{m+1}$ is A-orthogonal to $\boldsymbol{g}^m$, $(\boldsymbol{g}^{m+1}, A\boldsymbol{g}^m) = 0$, i.e.,

$$\beta^m = \frac{(\boldsymbol{\xi}^{m+1}, K\boldsymbol{\xi}^{m+1})}{(\boldsymbol{\xi}^m, K\boldsymbol{\xi}^m)}. \tag{2.45}$$

Hestenes and Stiefel (1952) showed that the vectors $\boldsymbol{\xi}^0, \boldsymbol{\xi}^1, \ldots$, and $\boldsymbol{g}^0, \boldsymbol{g}^1$, $\ldots$, generated by the above procedure satisfy the following relations:

$$(\boldsymbol{\xi}^i, K\boldsymbol{\xi}^j) = 0 \qquad \text{for } i \neq j, \tag{2.46}$$
$$(\boldsymbol{g}^i, A\boldsymbol{g}^j) = 0 \qquad \text{for } i \neq j, \tag{2.47}$$
$$(\boldsymbol{\xi}^i, A\boldsymbol{g}^j) = 0 \qquad \text{for } i \neq j \text{ and } i \neq j+1. \tag{2.48}$$

2.6 Conjugate-Gradient Acceleration

Since the spectral property of coefficient matrix A affects the convergence of the CG method, we can accelerate the convergence by substituting

$$\tilde{F}(x) = \frac{1}{2}(x, KAx) - (x, Kb) \qquad (2.49)$$

for (2.35) with an appropriate matrix K. For example, if K is similar to A^{-1}, the spectral property of matrix KA can contribute to rapid convergence. In the case of the Poisson equation, since the Green's function of the Laplacian is $1/|r|$, the inverse matrix of the discrete Laplacian is approximated as

$$K_{ij} = \frac{1}{|r_i - r_j|}, \qquad (2.50)$$

where r_i is the coordinate of the i-th grid point. The matrix K in (2.50), however, is not sparse; the operation regarding the matrix K is computationally demanding. Therefore we truncate it as follows:

$$K_{ij} = \sum_k \gamma_{1,k} \exp(-\gamma_{2,k}|r_i - r_j|^2) \qquad (2.51)$$

with certain parameters $\gamma_{1,k}$ and $\gamma_{2,k}$. Figure 2.2 shows the convergence in solving the Poisson equation. The computational model is the same as that described in Section 2.2 and the grid spacing is 0.3 a.u. which is ordinarily employed in practical real-space finite-difference calculations. One can recognize that the preconditioned CG method reduces the number of iterations compared with the CG method without preconditioning.

2.7 Multigrid Method

In the case of using simple iterations such as Jacobi and Gauss–Seidel procedures, although the high-frequency components of error are efficiently removed, the low-frequency components of error remain. The reason is that the representation of matrix A is near-local in real space which causes the stalling process of iterative solvers. The multigrid technique was developed based on the following key steps, in order to overcome this inherent difficulty in real-space methods [Brandt (1977); Brandt *et al.* (1983); Hackbusch (1985); Briggs (1987); Wesseling (1991); Beck (1999, 2000)]. In this section, the key procedure of the multigrid technique is briefly introduced.

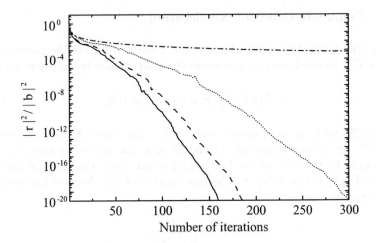

Fig. 2.2   Convergence characteristics. The dashed and single-dotted (dotted) curve is the result obtained using the steepest-descent method (CG method without preconditioning). The dashed curve is the result obtained with $\gamma_{1,1} = 0.8$, $\gamma_{2,1} = 30$, $\gamma_{1,2} = 0.2$, and $\gamma_{2,2} = 10$. The solid curve is the result obtained with $\gamma_{1,1} = 0.6$, $\gamma_{2,1} = 25$, $\gamma_{1,2} = 0.4$, and $\gamma_{2,2} = 8$.

Assume there are l levels of the grid spacing for the general case; each level is labeled by the index k, which has a value from 1 (the coarsest level) to l (the densest level). The Poisson equation to be solved on the k-th level problem is written as

$$A_k x_k = b_k + \tau_k, \tag{2.52}$$

where A_k, x_k, b_k, and τ_k are the finite-difference Laplacian, the current approximation of the Hartree potential, -4π times the charge density, and the defect correction on the k-th level, respectively. The initial x_k on a coarse level is obtained by applying the full-restriction operator I_{k+1}^{k} which performs a weighted local average of the dense level function to x_{k+1}:

$$x_k = I_{k+1}^{k} x_{k+1}. \tag{2.53}$$

Also, the defect correction τ_k is defined as

$$\tau_k = A_k I_{k+1}^{k} x_{k+1} - I_{k+1}^{k} A_{k+1} x_{k+1} + I_{k+1}^{k} \tau_{k+1}. \tag{2.54}$$

It is noted that the third term on the right-hand side of (2.54) is zero for the grid next-coarser to the dense level, since the defect correction is zero on the densest level l.

(a)

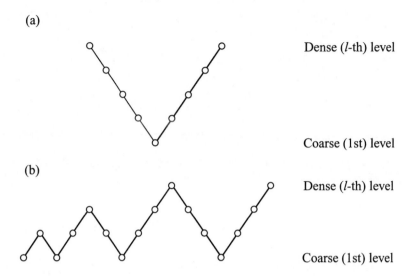

Dense (*l*-th) level

Coarse (1st) level

(b)

Dense (*l*-th) level

Coarse (1st) level

Fig. 2.3　Multigrid cycles. (a) V cycle. (b) Full-multigrid cycle. Iterations begin on the left side of the diagram.

The solver begins with initial iterations on the densest level to remove the high-frequency modes of error. The problem is passed to a coarser level to remove the low-frequency modes of error, e.g., $H = 2h$ where H and h are the grid spacings of the coarse level and the given level, respectively. The solver then returns to a dense level in order to correct the error caused by the interpolation from the coarse level. The correction equation for grid $k + 1$ is

$$x_{k+1} \leftarrow x_{k+1} + I_k^{k+1}(x_k - I_{k+1}^k x_{k+1}), \qquad (2.55)$$

where I_k^{k+1} is the interpolation operator. By performing these processes recursively through several levels [see Fig. 2.3 (a)], the errors of all wavelengths can be effectively removed; this cycle is called the V cycle. If iterations begin on the coarsest level, as shown in Fig. 2.3 (b), the procedure is a full-multigrid method. The advantage of the full multigrid is that a good initial approximation of the dense-grid function is obtained on the left side of the final V cycle.

Chapter 3

Minimization Procedures of the Energy Functional

Tomoya Ono and Kikuji Hirose

3.1 Introduction

The ground-state electronic structure can be obtained by minimizing the energy functional of (1.1) with the constraint of orthonormality of the wave functions:

$$\tilde{E}_{tot} = E_{tot} - 2 \sum_{ij} n_i \varepsilon_{ij} (\langle \psi_i | \psi_j \rangle - \delta_{ij}), \qquad (3.1)$$

where ε is the Lagrange multiplier, and n_i is the occupation number of each wave function and includes a k-point weight if there is more than one k-point. The Lagrange multipliers ε_{ii} ensure that wave functions remain normalized, while the Lagrange multipliers ε_{ij} $(i \neq j)$ ensure that wave functions remain orthogonal. When the orthogonality constraints are maintained by other procedures, the differential of $\tilde{E}_{tot}$ with respect to ψ_i^* is

$$\frac{\partial \tilde{E}_{tot}}{\partial \psi_i^*} = 2n_i (H\psi_i - \varepsilon_{ii}\psi_i), \qquad (3.2)$$

where H is the Kohn–Sham Hamiltonian.

As described in Section 2.5, the minimization technique considers a functional F of the vector x in which the gradient of the functional can be calculated using a gradient operator. In the case of the total energy calculation, the energy functional E_{tot} or the expectation value of the Kohn–Sham Hamiltonian, i.e., the eigenvalue, takes the place of the functional F, and the wave function ψ_i takes the place of the vector x.

3.2 Approach of Minimizing the Kohn–Sham Energy Functional

A conjugate-gradient iteration can be used to update a single band at a time [Teter *et al.* (1989)]. The steepest-descent direction ψ_i^m for a single band is given by

$$\psi_i^{(1)m} = -(H - \varepsilon_i)\psi_i^m, \tag{3.3}$$

where $\varepsilon_i = \langle \psi_i^m | H | \psi_i^m \rangle$, the superscript m indicates the iteration number, and the subscript i represents the band. In addition, the occupation number $2n_i$ can be disregarded since the conjugate vectors are normalized in a later procedure.

A Kohn–Sham-energy-functional minimization in which the wave functions are constrained to be orthogonal is different from the conjugate-gradient minimization employed to solve the Poisson equation. Since the steepest-descent vector must be orthogonal to all other bands, it is orthogonalized as

$$\psi_i^{(2)m} = \psi_i^{(1)m} - \sum_{j<i} \langle \psi_j | \psi_i^{(1)m} \rangle \psi_j. \tag{3.4}$$

In the formalism proposed by Teter *et al.* (1989), the preconditioning technique is employed to accelerate the convergence. In our own work, the preconditioning matrix K generated by the inverse matrix of the discrete Laplacian, as explained in Section 2.6, is employed. The other preconditioning matrices for the Kohn–Sham-energy-functional minimization proposed by Hoshi *et al.* (1995) and Seitsonen *et al.* (1995) are also effective. The preconditioned steepest-descent vector is

$$\psi_i^{(3)m} = K\psi_i^{(2)m}. \tag{3.5}$$

The preconditioned steepest-descent vector $\psi_i^{(3)m}$ is not orthogonal to all bands, and therefore is orthogonalized as

$$\psi_i^{(4)m} = \psi_i^{(3)m} - \langle \psi_i^m | \psi_i^{(3)m} \rangle \psi_i^m - \sum_{j \neq i} \langle \psi_j | \psi_i^{(3)m} \rangle \psi_j. \tag{3.6}$$

Then the preconditioned conjugate direction along which to minimize the functional is

$$\psi_i^{(5)m} = \psi_i^{(4)m} + \gamma^m \psi_i^{(5)m-1}, \tag{3.7}$$

where

$$\gamma^m = \frac{\langle \psi_i^{(4)m} | \psi_i^{(2)m} \rangle}{\langle \psi_i^{(4)m-1} | \psi_i^{(2)m-1} \rangle} \qquad (m > 1), \tag{3.8}$$

and $\gamma^1 = 0$. Here, the conjugate direction $\psi_i^{(5)m}$ is orthogonal to ψ_j, but not orthogonal to the present band ψ_i^m. The new conjugate direction is orthonormalized as

$$\psi_i^{(6)m} = \psi_i^{(5)m} - \langle \psi_i^m | \psi_i^{(5)m} \rangle \psi_i^m, \tag{3.9}$$

and

$$\psi_i^{(7)m} = \frac{\psi_i^{(6)m}}{\langle \psi_i^{(6)m} | \psi_i^{(6)m} \rangle^{1/2}}. \tag{3.10}$$

Then we add the conjugate vectors to the present band so as to minimize the functional:

$$\psi_i^{m+1} = \psi_i^m \cos\theta + \psi_i^{(7)m} \sin\theta. \tag{3.11}$$

Here, in order to maintain normality, the triangular functions are employed. Substituting (3.11) into (3.1), we obtain a good approximation of the energy functional over the entire range of θ:

$$\tilde{E}_{tot}(\theta) = E_{avg} + A_1 \cos(2\theta) + B_1 \sin(2\theta). \tag{3.12}$$

Three pieces of information are required to evaluate an approximate minimum of $\tilde{E}_{tot}$. In the case of $\theta = 0$, we can easily compute $\tilde{E}_{tot}$ since $\psi_i^{m+1} = \psi_i^m$. Differentiating $\tilde{E}_{tot}$ with respect to ψ_i^m and $\psi_i^{(7)m}$, and then differentiating them with respect to $\theta = 0$, we have

$$\left. \frac{\partial \tilde{E}_{tot}}{\partial \theta} \right|_{\theta=0} = 2n_i \langle \psi_i^{(7)m} | H | \psi_i^m \rangle + 2n_i \langle \psi_i^m | H | \psi_i^{(7)m} \rangle$$

$$= 4n_i \, \mathrm{Re}(\langle \psi_i^{(7)m} | H | \psi_i^m \rangle). \tag{3.13}$$

The differential of $\tilde{E}_{tot}$ in (3.12) at $\theta = 0$ is

$$\left. \frac{\partial \tilde{E}_{tot}}{\partial \theta} \right|_{\theta=0} = 2B_1. \tag{3.14}$$

From (3.13) and (3.14),

$$B_1 = 2n_i \, \mathrm{Re}(\langle \psi_i^{(7)m} | H | \psi_i^m \rangle). \tag{3.15}$$

If $\tilde{E}_{tot}$ at the point $\theta_1 = \pi/300$ is computed, the other two unknowns are calculated as

$$E_{avg} = \frac{\tilde{E}_{tot}(\theta_1) - \tilde{E}_{tot}(0)\cos(2\theta_1) - B_1\sin(2\theta_1)}{1 - \cos(2\theta_1)}, \qquad (3.16)$$

and

$$A_1 = \frac{\tilde{E}_{tot}(0) - \tilde{E}_{tot}(\theta_1) + B_1\sin(2\theta_1)}{1 - \cos(2\theta_1)}. \qquad (3.17)$$

Once the parameters E_{avg}, A_1, and B_1 have been evaluated, the value of θ which minimizes $\tilde{E}_{tot}$ can be calculated. The stationary point is

$$\theta_{min} = \frac{1}{2}\tan^{-1}\frac{B_1}{A_1}, \qquad (3.18)$$

and the new wave function used in the next iteration is

$$\psi_i^{m+1} = \psi_i^m \cos\theta_{min} + \psi_i^{(7)m} \sin\theta_{min}. \qquad (3.19)$$

The calculation of an analytic second derivative of the $\tilde{E}_{tot}$ at $\theta = 0$ provides a more elegant way of determining the θ_{min}. For reviews of this scheme, see Payne *et al.* (1992).

3.3 Approach of Minimizing the Kohn–Sham Eigenvalues

The conjugate-gradient method of Teter *et al.* (1989) is effective for the electronic structure calculation for insulators and semiconductors. However, this approach fails miserably in the calculations for metallic systems since the total energy cannot be converged [Needels *et al.* (1992)]. In metallic systems, a bunch of single-particle states nest near the Fermi level, and when these states cross the Fermi level during the iterations, discontinuous changes in the orbital occupations occur because this method cannot treat unoccupied states strictly. Bylander *et al.* (1990) proposed a variation of the conjugate-gradient method by which unoccupied states can be computed in the same manner as occupied states.

Their variational method is doubly iterative in which one iterative method is used to improve each approximate eigenfunction and then the other iterative method is implemented for diagonalizing a matrix of the improved approximate eigenfunctions in order to obtain starting eigenfunctions for the next iteration which begins after the charge density has been updated. The procedure for obtaining the conjugate vector is the same as the method of Teter *et al.* (1989). The wave function is updated as

$$\psi_i^{(8)m} = \alpha\psi_i^m - \beta\psi_i^{(7)m}, \tag{3.20}$$

where $\alpha^*\alpha + \beta^*\beta = 1$. These coefficients are determined by diagonalizing the 2×2 matrix of H between ψ_i^m and $\psi_i^{(7)m}$ to minimize the eigenvalue of H. These iterations are performed for each band, and then the Ritz projection is implemented in order to improve the occupied subspace by making all residuals orthogonal to that subspace. The Ritz projection can be written as

$$\mathcal{H}c_i = \varepsilon_i c_i, \tag{3.21}$$

and

$$\psi_i^{m+1} = \sum_{j=1}^{M} c_{j,i}\psi_j^{(8)m}, \tag{3.22}$$

where $\mathcal{H}$ is the $M \times M$ (M is the number of wave functions)-dimensional matrix between $\{\psi_i^{(8)m}\}$, and c_i is the coefficient vector whose elements are $c_{j,i}$.

After computing ψ_i^{m+1}, the charge density ρ^m, which is the output of the m-th iteration, is mixed with the charge density ρ_{old}, which was used to evaluate the potential of the m-th iteration, for the purpose of starting the next iteration:

$$\rho_{new} = (1 - \zeta)\rho_{old} + \zeta\rho^m, \tag{3.23}$$

where ζ is a mixing ratio. The self-consistency of the Kohn–Sham equation is assured when the difference between ρ_{old} and ρ^m becomes small.

3.4 Multigrid Method for Minimizing the Kohn–Sham Energy Functional

Brandt *et al.* (1983) extended the multigrid algorithm (see Section 2.7) to the problem of minimizing the total energy of (3.1). The basic procedure is similar to that in Poisson problems: the Laplacian operator A_k in (2.52) is now replaced by the real-space Kohn–Sham Hamiltonian minus the eigenvalue of the Hamiltonian ε_i, the vector b is a zero vector and x is a wave function ψ_i^m. The functional (3.1) is minimized according to the steepest-descent direction by the Gauss–Seidel or other alternative method. The following complexities were introduced: (i) calculation of multiple wave functions, (ii) orthonormality of the wave functions, (iii) computation of eigenvalues, and (iv) Ritz projection.

The orthonormalization and eigenvalues are required only on the coarsest level where the computational expense is small. In the case of orthonormalization, a direct Gram–Schmidt method is not applicable on coarse levels; if the exact grid solution is restricted to the coarse levels, the resulting wave functions are no longer orthonormal. Therefore, to satisfy the zero correction at convergence condition, a coarse grid matrix equation for the constraints is given:

$$(\psi_{i,k}^m, I_1^k \psi_{j,1}^m) = (I_1^k \psi_{i,1}^m, I_1^k \psi_{j,1}^m), \tag{3.24}$$

where $\psi_{i,k}^m$ is the wave function at the m-th iteration, i-th band, and k-th multigrid level and $I_1^k = I_{k-1}^k \times I_{k-2}^{k-1} \cdots I_2^3 \times I_1^2$. The solution requires inversion of a $M \times M$ matrix, where M is the number of wave functions. The inversion can be accomplished by a direct matrix method since M is quite small compared with the number of grid points. The eigenvalue at the k-th multigrid level is written as

$$\varepsilon_i = \frac{(H_k \psi_{i,k}^m - \tau_{i,k}, \psi_{j,1}^m)}{(\psi_{i,k}^m, \psi_{i,k}^m)}, \tag{3.25}$$

where H_k is the real-space Kohn–Sham Hamiltonian on the k-th level and $\tau_{i,k}$ is the defect correction for the i-th band on the k-th level. Relaxation steps are performed on each level through Gauss–Seidel iterations. Finally, Brandt *et al.* (1983) added a Ritz projection performed at the conclusion of each V cycle in the full-multigrid solver.

There are some improved versions of the multigrid technique for the problem of minimizing the total energy. Beck (2000) proposed updating the wave functions simultaneously. In addition, Costiner and Ta'asan (1995) transferred the Ritz projection step to coarse grids and added a backrotation in order to prevent rotations of the solutions in a subspace of equal or close eigenvalues of the Hamiltonian. They also developed an adaptive clustering algorithm for handling groups of wave functions with near eigenvalues.

3.5 Fermi-Broadening Technique

In the Kohn–Sham formulation of the density-functional theory, the energy functional of a system of electrons is minimum at the point of the ground-state energy with respect to the variation of an arbitrary set of single-particle orthonormal wave functions $\{\psi_i\}$ and occupation numbers $\{n_i\}$.

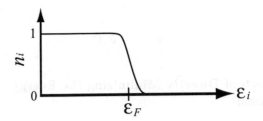

Fig. 3.1   Fermi-distribution function.

The Janak theorem [Janak (1978)] implies that the occupation numbers of the ground state at zero temperature are chosen such that $n_i = 1$ for the states of Kohn–Sham eigenvalues below the Fermi level, and $n_i = 0$ for those above the Fermi level. In calculating electronic wave functions, however, constraining the occupation numbers to be integers can frequently cause a convergence difficulty for metallic systems, because a bunch of single-particle states nest in the energy range including the Fermi level and some states move through the Fermi level during the course of self-consistent calculations, i.e., level crossing happens to occur [Harris (1984); Fernando *et al.* (1989)].

To ensure the convergence of the total energy, the fractional occupation numbers are introduced by using the following Fermi-distribution (FD) function with fictitious temperature T [Fernando *et al.* (1989); Weinert and Davenport (1992); Wentzcovitch *et al.* (1992)] (see Fig. 3.1):

$$n_i = \frac{1}{1 + \exp[(\varepsilon_i - \varepsilon_F)/k_B T]}. \tag{3.26}$$

Here, k_B is the Boltzmann constant, ε_F is the Fermi level and ε_i is an eigenvalue of the Kohn–Sham Hamiltonian. This technique is efficiently used in the above-mentioned minimization procedures for metallic systems.

In the case of employing fractional occupation numbers, the thermodynamic free energy

$$\Omega = \tilde{E}_{tot} - TS \tag{3.27}$$

is minimized instead of the electronic total energy, where S denotes the entropy associated with the occupation numbers,

$$S = k_B \sum_i [n_i \ln n_i + (1 - n_i) \ln(1 - n_i)]. \tag{3.28}$$

Then the property of being stationary with respect to n_i makes the gradient of the free energy exactly equal to the Hellmann–Feynman force acting on the s-th ion

$$\mathbf{F}_s = -\nabla_s \Omega. \tag{3.29}$$

3.6 Approach of Directly Minimizing the Energy Functional

In the preceding section, we have seen one popular approach to prevent the numerical difficulty due to the level crossing in which the fractional occupation numbers are introduced by using the FD function with broadening temperature. In principle, however, no physical significance is given to temperature or entropy in the context of the zero-temperature Kohn–Sham scheme. It is even more serious in practice that theoretical results depend on the chosen values of the fictitious broadening temperature [Wagner *et al.* (1998); Mehl (2000)], and that one can not always eliminate this dependence by lowering the temperature down to zero in metallic systems, since the self-consistent cycle frequently enters into the endless loop at low temperature due to the restarting of the level crossing, typical examples of which are found in Pederson and Jackson (1991), and Wentzcovitch *et al.* (1992).

In this section, we demonstrate that the direct minimization (DM) of the energy functional proposed by Mauri, Galli and Car (MGC) [Mauri *et al.* (1993); Mauri and Galli (1994)] can completely avoid the numerical difficulty caused by the level crossing, and can give satisfactorily the self-consistent solutions of the Kohn–Sham equation without usage of conventional self-consistent field techniques: not only the manifold of the correct electronic wave functions but also the proper occupation numbers are numerically determined as output quantities.

The MGC energy functional for an N-electron system is written as

$$E[\{\phi\}, \eta] = 2 \sum_{i,j}^{M} Q_{ij} \langle \phi_j | - \frac{1}{2} \nabla^2 | \phi_i \rangle + F[\rho]$$

$$+ \eta \left\{ N - \int \rho(\mathbf{r}) d\mathbf{r} \right\} - \sum_s Z_s v_f(\mathbf{R}^s) + \gamma_E, \tag{3.30}$$

where $\{\phi_i\}$ is an arbitrary set of M linearly independent *overlapping* wave functions, which are assumed here to be real functions for simplicity, and M is taken to be not smaller than the number of the occupied states for each spin. $F[\rho]$ is the sum of the external, Hartree, and exchange-correlation potential energy functionals, η is the electronic chemical potential, γ_E is the Coulomb energy associated with interactions among the nuclei, v_f is the external-electric-field potential, Z_s is the valence of the s-th atom, Q

is an $(M \times M)$ matrix: $Q_{ij} = 2\delta_{ij} - S_{ij}$, and S_{ij} is the overlap matrix: $S_{ij} = \langle \phi_i | \phi_j \rangle$. The charge density is defined as

$$\rho(\mathbf{r}) = 2 \sum_{i,j}^{M} Q_{ij} \phi_j(\mathbf{r}) \phi_i(\mathbf{r}). \tag{3.31}$$

The factor 2 accounts for the same contribution from up and down spin electrons. The form of the energy functional (3.30) was originally introduced for computation with linear system-size scaling $O(N)$, and each wave function ϕ_i was approximated to be a Wannier-like function localized in an appropriate region of space, i.e., a localized orbital, to reduce the amount of computation. Here, we adhere to ordinary *extended* wave functions, being free from the errors caused by the localization of wave functions [Hirose and Ono (2001)].

$E[\{\phi\}, \eta]$ is minimized by a steepest-descent or conjugate-gradient algorithm *without constraint of the orthonormalization of wave functions*. The derivative of the functional with respect to the function ϕ_i is required for the minimization, which is given by

$$\frac{\delta E[\{\phi\}, \eta]}{\delta \phi_i} = 4 \sum_{j}^{M} \Big[(\hat{H}[\{\phi\}] - \eta) |\phi_j\rangle Q_{ji}$$
$$- |\phi_j\rangle \langle \phi_j | (\hat{H}[\{\phi\}] - \eta) |\phi_i\rangle \Big], \tag{3.32}$$

where $\hat{H}[\{\phi\}]$ is the Kohn–Sham Hamiltonian.

Theorem 3.1 *The set of the single-particle wave functions minimizing the Kohn–Sham energy functional (1.1),*

$$|\phi_i^0\rangle = a_i |\psi_i\rangle, \tag{3.33}$$

is a stationary point of $E[\{\phi\}, \eta]$, when the set of the coefficients $\{a_i\}$ is such that $|a_i| = 1$ for $\varepsilon_i < \eta$, $0 \leq |a_i| \leq 1$ for $\varepsilon_i = \eta$, and $a_i = 0$ for $\varepsilon_i > \eta$. Here, $|\psi_i\rangle$ is the normalized eigenfunction of $\hat{H}$ with the eigenvalue ε_i, i.e., $\hat{H}[\{\phi_i^0\}]|\psi_i\rangle = \varepsilon_i |\psi_i\rangle$.

Proof. Substituting (3.33) to (3.32), we have

$$\frac{\delta E[\{\phi\}, \eta]}{\delta \phi_i} \bigg|_{\{\phi_i\} = \{\phi_i^0\}} = 8a_i (1 - |a_i|^2)(\varepsilon_i - \eta) |\psi_i\rangle$$
$$= 0. \tag{3.34}$$

Thus, $\{\phi_i^0\}$ is a stationary point of $E[\{\phi\}, \eta]$. □

Theorem 3.2   *The MGC energy functional (3.30) is identical to the functional in the standard form of (1.1) except that the occupation number n_i varies in the range of $-\infty < n_i \leq 1$.*

Proof.   A set of $\{\phi_i\}$ in (3.30) is transformed to an orthonormal set $\{\tilde{\psi}_i\}$ using an orthogonalization algorithm, e.g., the Gram–Schmidt method, and then ϕ_i is represented as $\phi_i = \sum_l^M c_{li}\tilde{\psi}_l$. Substituting this expansion into (3.30) and (3.31), we obtain for the energy functional

$$E[\{\tilde{\psi}\}, \eta] = 2 \sum_{k,l}^M P_{kl} \langle \tilde{\psi}_l | - \frac{1}{2}\nabla^2 | \tilde{\psi}_k \rangle$$

$$+ F[\rho] + \eta \left\{ N - \int \rho(\mathbf{r}) d\mathbf{r} \right\}, \tag{3.35}$$

and a similar expression for the charge density, where

$$P_{kl} = 2 \sum_{i,j}^M Q_{ij} c_{k,i} c_{l,j}$$

$$= 4T_{kl} - 2 \sum_m^M T_{km} T_{ml}, \tag{3.36}$$

with $T_{kl} = \sum_i^M c_{ki} c_{li}$. Here, the matrix T, of which the element of the k row and the l column is T_{kl}, is a non-negative definite Hermitian matrix, which can be diagonalized by a unitary matrix U as $(U^\dagger T U)_{kl} = \delta_{kl}\lambda_k$ with λ_k being a non-negative eigenvalue of T. Hence,

$$(U^\dagger P U)_{kl} = 2\delta_{kl}(2 - \lambda_k)\lambda_k. \tag{3.37}$$

Defining ψ_i and n_i as $\psi_i = \sum_j^M U_{ji}\tilde{\psi}_j$ and $n_i = (2 - \lambda_i)\lambda_i$, we obtain (1.1) from (3.35), and the inequality $-\infty < n_i \leq 1$.   □

This proof gives the way how to define and calculate the occupation numbers $\{n_i\}$ within the MGC formalism. One can now recognize that minimizing the MGC energy functional (3.30) with respect to $\{\phi_i\}$ without imposing the normalization condition is equivalent to treating the occupation numbers $\{n_i\}$ as *fractional* occupation variables in the minimization of the functional (1.1). It is also noteworthy that since the occupation numbers defined above do not exceed one, the Pauli principle automatically works to prevent more than one electron from falling into each single-particle level of each spin in the course of the minimization of the energy functional (3.30). Kim *et al.* (1995) showed within a non-self-consistent scheme that the occupation number of the single-particle level below (above) the Fermi level is an asymptote toward one (zero) in the vicinity of the stationary ground-state point of the MGC energy functional. See Fig. 1 in Kim *et al.* (1995), in which the variable a_i is connected with the occupation number n_i as $n_i = (2 - a_i^2)a_i^2$. Following this argument, one can easily see that as long as initial wave functions for iterations are chosen to be close to the ground-state solution, the variation range of the occupation number is bounded to $0 \leq n_i \leq 1$ and unphysical negative occupation gives rise to no problem in practical calculations. In any case, one should not worry when some occupation numbers are temporarily negative during the minimization process before converging to the stationary point because these occupations do not have a direct physical meaning.

The overall computational scaling in the present DM procedure combined with the real-space finite-difference method amounts to $O(M^2 N_{xyz})$ operations in the calculations of the energy functional (3.30) and its derivative (3.32), since the ϕ_is are now assumed to be *extended* wave functions. It is noted that $O(M^2 N_{xyz})$ is equal to the scaling order in the orthogonalization of wave functions indispensable in the conventional approach of solving the Kohn–Sham equation by an iterative algorithm, and that the dominant scaling in the calculations of the energy functional and its derivative based on a plane-wave basis set is also $O(M^2 N_{basis})$ with N_{basis} being the number of plane-wave basis functions, when advantage is taken of the fast Fourier transform technique. Moreover, when a system is assumed to be described in terms of localized wave functions, the present DM procedure makes it possible to perform the calculation with linear system-size scaling $O(N)$ (see Section 3.8).

3.7 Illustration for Direct Minimization Efficiency

In order to illustrate the utility of the DM method and the difficulty in the usage of the FD function, we evaluate the electronic structures of C_2 and Si_2 molecules. Computational conditions are as follows: the nine-point finite-difference formula, i.e., $N_f = 4$ in (1.5), for the derivative arising from the kinetic-energy operator is employed, and the dense-grid spacing is fixed at $h_{dens} = h/3$, where h denotes the coarse-grid spacing (see Section 4.1). We use the norm-conserving pseudopotential [Kobayashi (1999a)] of Troullier and Martins (1991) in a separable form of Kleinman and Bylander [see Eq. (1.23)].

Figures 3.2 and 3.3 show the results of the application to the carbon dimer as a demonstration of the potential power that the DM method can correctly evaluate both the manifold of the electronic wave functions and the occupation numbers. The carbon dimer is one of the most suitable examples for this purpose, because a number of investigations [Dunlap (1984, 1988); Harris (1984); Pederson and Jackson (1991); Chelikowsky *et al.* (1994a, 1994b)] already showed that the mathematical minimum of the energy functional is with fractional occupations in a certain range of the interatomic distance, e.g., Pederson and Jackson (1991) found that the fractionally occupied states appear over the range of interatomic distance $d = 2.4 - 3.7$ a.u. and the Kohn–Sham eigenvalues of these states are degenerate at the Fermi level. There has been much discussion about the physical significance of fractionally occupied states, e.g., see Harris (1984). From a pragmatic point of view, however, fractional occupation numbers provide the smoothest path to the ground-state geometry as seen in Fig. 3.2(a). In Figs. 3.2 and 3.3, we took the grid spacing $h = 0.33$ a.u. in a cell of $16.0 \times 16.0 \times 16.0$ a.u. under the nonperiodic boundary condition of vanishing wave function, and set the number of electrons $N = 8$ and the number of wave functions $M = 5$. Exchange-correlation effects were treated with the local-density approximation [Perdew and Zunger (1981)] according to Pederson and Jackson (1991). Figure 3.2 illustrates the ground-state total energy, the Kohn–Sham eigenvalues of the $1\pi_u$ and $2\sigma_g$ states, and the occupation number of the $1\pi_u$ state as a function of the interatomic distance. The result of the brute-force method [Dunlap (1984, 1988)] implementing the manual determination of the occupation numbers is also depicted here in confirmation of the accuracy of our DM method. The results obtained from our DM method accord with those from the other method all over the range of the atomic distance, and we can confirm the accuracy and efficiency of our DM method. The calculated equilibrium bond length of the ground state geometry is $R = 2.34$ a.u., being out of the degenerate range, which is in agreement with the experimental value 2.35 a.u. In Fig. 3.3, we plot the occupation numbers of the states at $d = 2.30$ a.u. and 2.50 a.u. as a function of the number of minimization iterations, where the electronic structure at $R = 2.40$ a.u. is used as the starting point. One can see that the iterative process is markedly stable and our DM procedure gives good convergence of the occupation numbers.

We next give an example that the FD method including a fictitious broadening temperature leads to an incorrect ground-state geometry in molecular-dynamics simulations. Figures 3.4(a) and 3.4(b) show the calculated results of the free energy and the force on each atom versus the atomic separation for the silicon dimer, respectively. Here, the grid resolution $h = 0.50$ a.u., a cell of $24.0 \times 24.0 \times 24.0$ a.u. under the nonperiodic boundary condition and the local-spin-density approximation [Perdew and Zunger (1981)] for the exchange-correlation potential. The experiments [Huber and Herzberg (1979)] proved that the energy difference between two molecular configurations (i) $(1\sigma_g)^2(1\sigma_u)^2(1\pi_u)^4$ and (ii) $(1\sigma_g)^2(1\sigma_u)^2(2\sigma_g)^2(1\pi_u)^2$

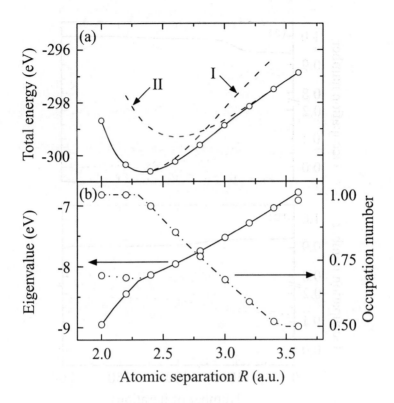

Fig. 3.2 C$_2$ adiabatic-potential curve, eigenvalues of $1\pi_u$ and $2\sigma_g$ states, $1\pi_u$ occupancy as a function of atomic separation R. In (a), the solid adiabatic-potential curve is the result with the fractional occupancy by our DM method, and the dashed curves I and II are those with the occupation of the σ_g state constrained to be 0 and 1, respectively. In (b), the solid (dotted) curve represents the eigenvalue of $1\pi_u$ ($2\sigma_g$) state, and the dash-dotted curve shows the occupation number of $1\pi_u$ state. Circles correspond to data obtained by the brute-force method. Data taken from Hirose and Ono (2001).

which form the double minimum is quite small, the equilibrium bond length is 4.07 a.u. for configuration (i) and 4.25 a.u. for (ii), and the ground state with the lowest total energy is the latter. Some theoretical analyses have been carried out to examine the situation [Northrup *et al.* (1983); Dunlap (1984, 1988); Chelikowsky *et al.* (1994a, 1994b)]. As seen in Fig. 3.4, the DM procedure can yield results that are in good agreement with the empirical data. On the contrary, the conventional FD method is unable to search out the global minimum configuration (ii) in molecular-dynamics iterations at a broadening temperature $T \geq 1$ mH. At a low temperature $T \simeq 0.5$ mH, the FD method leads to a mistake of regarding the configuration (i)

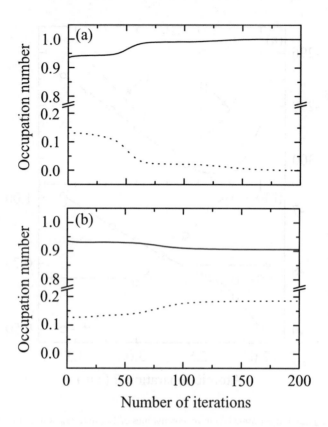

Fig. 3.3 The occupation number $n_i \equiv (2 - \lambda_i)\lambda_i$ as a function of the number of conjugate-gradient iterations. [For the definition of λ_i, see the text preceding (3.37).] The solid (dotted) curve represents the occupation numbers of the $1\pi_u$ ($2\sigma_g$) state. The curves in (a) and (b) are the results in the cases of the atomic separation $R = 2.30$ a.u. and 2.50 a.u., respectively, and each run starts from the electronic structure at $R = 2.40$ a.u. Reprinted with permission from Hirose and Ono (2001). © 2001 American Physical Society.

as the ground state with the lowest total energy, although one can find the minimum configuration (ii) by the FD calculations sweeping near the configuration (ii).

We have demonstrated that the DM procedure can completely avoid the convergence difficulty caused by the level crossing and evaluate the ground-state electronic structure with a high degree of accuracy. This scheme requires no additional statistical parameter such as a broadening temperature. Since many of those who study the *ab initio* molecular dynamics for metallic systems by using the FD broadening frequently worry about

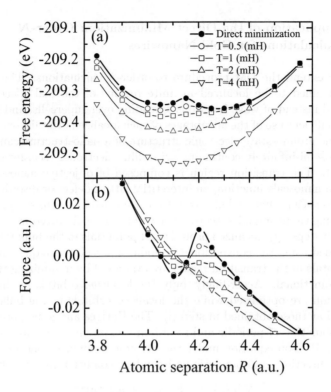

Fig. 3.4   (a) Si₂ free energy and (b) force on each Si atom as a function of atomic separation R obtained by the DM method and the conventional FD method at different broadening temperatures T. The calculated points are fit to spline-interpolated curves as a guide to the eye. The free energy in the DM method is identical to the total energy. In (b), the force in the FD method is defined as the derivative of the free energy with respect to the atomic position, while the force in the DM method is the true Hellmann–Feynman force defined as the derivative of the total energy. Reprinted with permission from Hirose and Ono (2001). © 2001 American Physical Society.

entering into the endless loop of the self-consistent cycle as the broadening temperature goes to zero and have a question for the reliability of the calculation at a finite value of the broadening temperature, this method would be a great blessing to them. For further details of the DM method, see Hirose and Ono (2001).

3.8 Application of the Direct Minimization: Order-N Calculations for Gold Nanowires

Here we extend the DM procedure to order-N calculations. The orbitals ϕ_i constrained to be localized in finite regions of the real space can be employed instead of wave functions spreading over supercells, and then, by making efficient use of the advantages of such localized orbitals, one can calculate the ground-state electronic structure of a nanostructure sandwiched between semi-infinitely continuing crystalline electrodes (see, for example, Fig. 3.5). The transition region is composed of objective nanostructures such as a nanoscale junction, an interstitial lattice defect or disorder, an interface between different bulks or a tunnel junction in a tip-sample system of scanning tunneling microscopy. The computational procedure consists of two main steps. (i) Localized orbitals and potentials in the bulk regions are determined using the conventional periodic supercell technique. (ii) Localized orbitals in the transition region are evaluated by minimizing the MGC energy functional. Accordingly, only the localized orbitals in the transition region are optimized while the localized orbitals in the bulk regions are fixed at those obtained in step (i). The Hartree and ionic potentials in the combination of periodic and semi-infinite system are computed by the following Poisson equation under the periodic boundary conditions in the x and y directions and nonperiodic boundary condition in the z direction:

$$\nabla^2 v(\boldsymbol{r}) = -4\pi(\rho(\boldsymbol{r}) + \rho_{ion}(\boldsymbol{r})), \tag{3.38}$$

where $v(\boldsymbol{r})$ is the sum of the Hartree and ionic potentials and $\rho_{ion}(\boldsymbol{r})$ is the charge density of ions. In order to set up the simultaneous equations for solving the Poisson equation, the boundary values of $v(0)$ and $v(N_z h_z + h_z)$ are required, where h_z and N_z are the grid spacing and the number of the grid point in the z direction, respectively. We adopt the potential $v(\boldsymbol{r})$ at the boundary calculated in step (i).

For demonstration of the efficiency of this method, we present test calculations on a single-row gold wire attached to Au(001) electrodes. First, the ground-state electronic structure of the wire is calculated using the present DM procedure, and then its electronic conductance is evaluated to confirm the accuracy of the electronic structure calculations. For the conductance calculation, a wave function infinitely extending over the entire system is determined by the overbridging boundary-matching method (see Chapter 6). The conductance G at the zero-bias limit is given by the Landauer–Büttiker formula of (6.6) [Büttiker *et al.* (1985)]. Figure 3.5 shows the calculation model employed here. The transition region consists of the single-row gold wire, pyramidal bases, and several Au(001) planes of electrodes. Computational conditions are as follows:

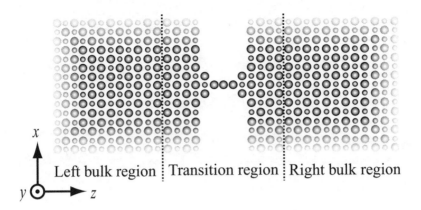

Fig. 3.5 Schematic representation of the gold wire system for $N_{atom} = 3$ with the transition region suspended between the semi-infinitely continuing bulk regions. The large and small spheres represent the atoms on and below the cross section, respectively.

The grid spacing is taken to be 0.68 a.u., which corresponds to a cut-off energy of 22 Ry in the conventional plane wave approach. We employ the central finite-difference formula, i.e., $N_f = 1$ in (1.5), for the derivative arising from the kinetic-energy operator and the local pseudopotential for ionic potential of atomic nuclei [Hamann *et al.* (1979); Bachelet *et al.* (1982)]. Exchange-correlation interaction between electrons is treated by the local density approximation [Perdew and Zunger (1981)]. The side lengths of the supercell in the x and y directions are $2a_0$, where a_0 (= 7.71 a.u.) is the lattice constant of the gold crystal. The number of atoms composing of the wire, N_{atom}, is varied from three to five. The size of the localization region for the localized orbitals is set to be 12.0 a.u. and the number of localized orbitals $M=176+N_{atom}$.

So far, it has been demonstrated experimentally that the single-row gold wire exhibits a conductance of ~ 1 G_0 $(= 2e^2/h)$ [Rubio *et al.* (1996); Ohnishi *et al.* (1998); Yanson *et al.* (1998); Smit *et al.* (2003)], where e is the electron charge and h is Plank's constant, and in addition, some theoretical studies on the conductance of gold wires observed the quantized conductance of ~ 1 G_0 [Okamoto and Takayanagi (1999); Brandbyge *et al.* (1999); Emberly and Kirczenow (1999); Fujimoto and Hirose (2003b)]. We show in Fig. 3.6 the conductance as the function of the number of atoms N_{atom} in the atomic wire. The conductance obtained by the present method is ~ 1 G_0. The calculated results are consistent with the experimental data and other theoretical data, which supports the applicability of this method. Further details of this simulation are given in Sasaki *et al.* (2004) and elsewhere.

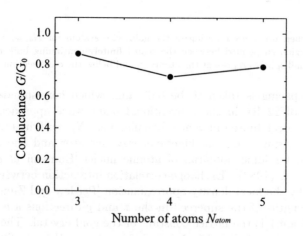

Fig. 3.6   Conductance of the atomic wires in units of the conductance quantum (G_0) as a function of the number of gold atoms in the wire. Data taken from Sasaki *et al.* (2004).

Chapter 4

Timesaving Double-Grid Technique

Tomoya Ono and Kikuji Hirose

4.1 Essential Feature of the Timesaving Double-Grid Technique

There is a disadvantage in the real-space finite-difference method, which is not found in conventional methods using basis functions: the total energy vibrates unphysically depending on the relative position of the grid points and the nucleus. This is caused by a deterioration of the accuracy of the inner product between pseudopotentials and wave functions or electron density distribution performed on discrete grid points. This problem can be avoided by reducing the grid spacing; however, this is not a practical solution because it requires a substantial increase in the computational effort. With this as background, the timesaving double-grid technique [Ono and Hirose (1999)] is introduced here as a simple method of circumventing this problem. The double-grid technique improves the accuracy of the numerical integral for the inner product between pseudopotentials and wave functions or electron density distribution by assigning grids denser than conventional coarse grids in the vicinity of nuclei where the pseudopotentials sharply vary, i.e., inside the inner shell of the pseudopotentials (see Fig. 4.1). As shown in Fig. 4.2, the pseudopotential varies more rapidly than either the wave function or the electron density distribution. Consequently, to improve the accuracy of the inner product, the grid spacing must be denser. To perform numerical integration for the inner product using a dense grid, values of the functions on the dense grids are required. In general, pseudopotentials are given exactly in an analytical or numerically exact manner; however, the values of wave functions and electron density distribution are given only on coarse grids. Therefore, their values on a

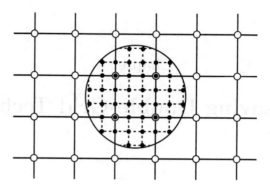

Fig. 4.1   Conceptual diagram of double grids. The ○ and ● represent coarse grids and dense grids, respectively. The circle shows the inner shell edge of the pseudopotential.

dense grid are evaluated by interpolating the values given on the coarse grid. There arises a concern regarding the increase in the computational effort due to interpolation; however, what is required here is not the product between the potential and the wave function or the electron density distribution on each dense-grid point, but instead, the integral values consisting of their sum. An advantage of the timesaving double-grid technique is that these integral values can be obtained at the same computational cost as when the calculation is carried out only on conventional coarse grids, as well as with the same accuracy as when it is performed on dense grids. The following is a proof of these assertions. For simplicity, we assume a one-dimensional case and use a linear interpolation.

The term of the ionic pseudopotential energy in the total energy is defined as [see Eqs. (1.25) and (1.28)]

$$
\begin{aligned}
E_{ne} &= E_{loc} + E_{nonloc} \\
&= \sum_s \Bigg[\int_\Omega v_{loc}^s(x - R_x^s)\rho(x)dx \\
&\quad + 2\sum_i^M n_i \sum_{lm} \frac{\left(\int_\Omega v_{lm}^s(x - R_x^s)\psi_i^*(x)dx\right)\left(\int_\Omega v_{lm}^{s*}(x - R_x^s)\psi_i(x)dx\right)}{\langle \psi_{lm}^{ps,s}|\hat{v}_l^s|\psi_{lm}^{ps,s}\rangle} \Bigg] \\
&= \sum_s \Bigg[\int_\Omega v_{loc}^{hard,s}(x - R_x^s)\rho(x)dx + \int_\Omega v_{loc}^{soft,s}(x - R_x^s)\rho(x)dx \\
&\quad + 2\sum_i^M n_i \sum_{lm} \frac{\left(\int_\Omega v_{lm}^s(x - R_x^s)\psi_i^*(x)dx\right)\left(\int_\Omega v_{lm}^{s*}(x - R_x^s)\psi_i(x)dx\right)}{\langle \psi_{lm}^{ps,s}|\hat{v}_l^s|\psi_{lm}^{ps,s}\rangle} \Bigg],
\end{aligned}
$$

$$(4.1)$$

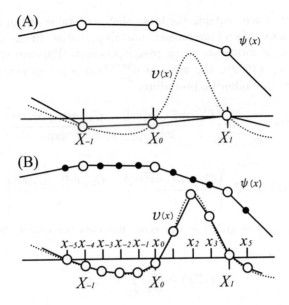

Fig. 4.2 Comparison of integral accuracies between calculations when the coarse grid alone is used and when the double-grid technique is used. (A) Conventional method. (B) Double-grid method. Here, X_J and x_j represent a coarse grid and a dense grid, respectively. The value for a wave function shown by (•) is obtained by interpolation. The schematic demonstrates that the double-grid technique can detect the detailed behavior of the pseudopotential which sharply changes.

where R_x^s, $v_{loc}^{hard,s}(x)$ and $v_{loc}^{soft,s}(x)$ represent the position of the nucleus, the local component of pseudopotential which varies rapidly around the nucleus (hard local part), and the local component of pseudopotential which changes gently (soft local part), respectively. The procedure to decompose the pseudopotential is explained below. In addition, $v_{lm}^s(x) = \hat{v}_l^s(x)\psi_{lm}^{ps,s}(x)$, where $\hat{v}_l^s(x)$ and $\psi_{lm}^{ps,s}(x)$ are the nonlocal parts of the pseudopotential and the pseudo-wave function used to prepare the pseudopotential, respectively. The numerical integrations related to $v_{loc}^{hard,s}(x)$ and $v_{lm}^s(x)$, which change sharply in the vicinity of nuclei, are performed on the dense grids using the timesaving double-grid technique. The integration related to $v_{loc}^{soft,s}(x)$, which changes gently, is implemented on the coarse grid. It is important that in the cases of $v_{loc}^{hard,s}(x)$ and $v_{lm}^s(x)$ to which the timesaving double-grid technique is applied, the integrated values must vanish outside the inner shell of the pseudopotential. Otherwise, in molecular-dynamics simulations, the calculation of the Pulay force is required, which is computationally demanding. Although the value of the nonlocal part is zero outside the inner shell, the value of the local part is not zero because it is a long-range Coulomb potential. In the case of using Bachelet *et al.*'s norm-conserving pseudopotential [Hamann *et al.* (1979); Bachelet *et al.* (1982)], the value of $v_{loc}^{hard,s}(x)$ is set to zero outside the inner shell by the following procedure:

$$v_{loc}^{hard,s}(x) = -Z^s \left[\frac{C_{1,s}\,\mathrm{erf}(\sqrt{\alpha_1^s}|x|)}{|x|} - \frac{C_{1,s}\,\mathrm{erf}(\sqrt{\alpha_2^s}|x|)}{|x|} \right], \qquad (4.2)$$

$$v_{loc}^{soft,s}(x) = -Z^s \left[\frac{C_{s,1}\,\mathrm{erf}(\sqrt{\alpha_2^s}|x|)}{|x|} + \frac{C_{2,s}\,\mathrm{erf}(\sqrt{\alpha_2^s}|x|)}{|x|} \right], \qquad (4.3)$$

where $\alpha_1^s \geq \alpha_2^s$ and $\mathrm{erf}(x)$ is the error function (or probability integral) defined by

$$\mathrm{erf}(x) = \frac{2}{\sqrt{\pi}} \int_0^x e^{-t^2}\,dt. \qquad (4.4)$$

Then, let us consider the integral of pseudopotential and the wave function or electron density distribution. First, the inner product of the hard local part and the electron density distribution on grids is given as

$$\int_{-r_c+R_s}^{r_c+R_s} v_{loc}^{hard,s}(x - R_x^s)\rho(x)dx \approx \sum_{j=-nN_{core}}^{nN_{core}} v_{loc,j}^{hard,s}\rho_j h_x^{dens}, \qquad (4.5)$$

where r_c is the inner shell radius of the pseudopotential function, $2N_{core}+1$ is the number of grid points within the inner shell, P_J is the value of the electron density at the coarse-grid point X_J, $2nN_{core} + 1$ is the number of dense-grid points, h_x^{dens} is the dense grid spacing, and h_x is the coarse grid spacing. When we employ the linear interpolation, the electron density distribution on a grid is approximated as

$$\rho_j = \frac{h_x - (x_j - X_J)}{h_x}P_J + \frac{h_x - (X_{J+1} - x_j)}{h_x}P_{J+1}. \qquad (4.6)$$

By substituting (4.6) into (4.5),

$$\sum_{j=-nN_{core}}^{nN_{core}} v_{loc,j}^{hard,s}\rho_j h_x^{dens} = \sum_{J=-N_{core}}^{N_{core}} w_{loc,J}^s P_J h_x$$

$$= 2\sum_i^M n_i \sum_{J=-N_{core}}^{N_{core}} \Psi_{i,J}^* w_{loc,J}^s \Psi_{i,J} h_x, \qquad (4.7)$$

where

$$w_{loc,J}^s = \sum_{k=-n}^{n} \frac{h_x - |x_{nJ+k} - X_J|}{nh_x} v_{loc,nJ+k}^{hard,s}, \tag{4.8}$$

and $\Psi_{i,J}$ is the wave function of the i-th state at the coarse-grid point X_J.

Next, the inner product of the nonlocal part and the wave function is similarly given by

$$\int_{-r_c+R_x^s}^{r_c+R_x^s} v_{lm}^{s*}(x - R_x^s)\psi_i(x)dx \approx \sum_{J=-N_{core}}^{N_{core}} w_{lm,J}^{s*}\Psi_{i,J}h_x, \tag{4.9}$$

$$w_{lm,J}^s = \sum_{k=-n}^{n} \frac{h_x - |x_{nJ+k} - X_J|}{nh_x} v_{lm,nJ+k}^s. \tag{4.10}$$

It should be noted that the integral on a dense grid is replaced by an integral on a coarse grid in (4.7) and (4.9). Furthermore, because the values of w_{loc}^s and w_{lm}^s are independent of the values of the wave function, the computations should be performed only once for each step in the molecular-dynamics simulations; the calculations are not necessary at every self-consistent iteration.

Consequently, although the integration is implemented on coarse grids in the timesaving double-grid technique, the inner products are evaluated with the same accuracy as in the case of using dense-grid points. Thus, by employing the timesaving double-grid technique, the computational costs can be considerably reduced while maintaining a high degree of accuracy.

4.2 Illustration of Double-Grid Efficiency

In this section, the efficiency of the timesaving double-grid technique is demonstrated by incorporating it into the formalism of the real-space finite-difference method. The following are the conditions applied to all calculations in this section.

(i) The ninth-order Lagrangian interpolation is used in the dense-grid interpolation.

(ii) The dense-grid spacing is set as $h_\mu^{dens} = h_\mu/3$. Here, h_μ ($\mu = x, y,$ and z) is the coarse-grid spacing.

(iii) The pseudopotential database NCPS97 [Kobayashi (1999a)] based on the norm-conserving pseudopotential devised by Troullier and Martins (1991) is used for the ionic potential.

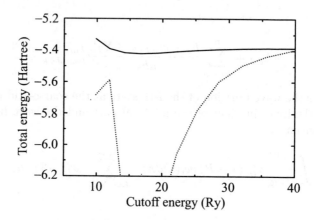

Fig. 4.3 Convergence of the total energy for the fluorine atom with respect to cutoff energy. The solid and dotted curves represent the results obtained with and without the double-grid technique, respectively.

(iv) The nine-point finite-difference formula, i.e., $N_f = 4$ [see Eq. (1.5)], is adopted for the differentiation of the wave function.
(v) Isolated boundary conditions are employed for all the directions.
(vi) The local-spin-density approximation [Perdew and Zunger (1981)] is used for the exchange-correlation potential.

First, the convergence of total energy for a fluorine atom as a function of the cutoff energy is examined. Because the p-orbital component of the pseudopotential in the nonlocal part oscillates in the vicinity of the nucleus, first-row elements are the most difficult atoms to treat by the real-space finite-difference method. The fluorine atom is placed in the center of the neighboring grid points for all x, y and z directions. The cutoff energy of the wave function is defined as $E_c^{coars} \equiv (\pi/h_\mu)^2$ Ry to correspond to the grid spacing used in the fast Fourier transform of the plane-wave expansion scheme [Gygi and Galli (1995)]. Figure 4.3 shows the convergence of total energy with respect to the cutoff energy. Without the use of the double-grid technique, the total energy does not converge even when the cutoff energy is increased to 33 Ry. On the other hand, when the double-grid technique is employed, the total energy converges even with a small cutoff energy; the obtained total energies converge rapidly and monotonically as the cutoff energy increases.

Next, the variation of the total energy is explored when the nucleus of a copper atom is displaced between adjacent grids. The coarse-grid spacing is set at 0.27 a.u. Because the pseudopotential of a transition-metal atom

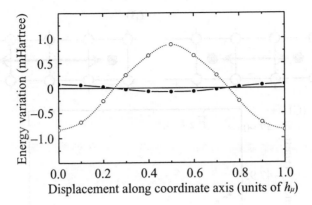

Fig. 4.4 Change in the total energy when a copper atom is moved between two grids. The coarse-grid spacing h_μ is 0.27 a.u. The solid and dotted curves represent the results obtained with and without the double-grid technique, respectively.

varies rapidly within the inner shell, the calculation of a model containing a transition-metal atom requires a high cutoff energy. Particularly in the calculation using a grid, such as the real-space finite-difference method, when grid spacing is coarse, the total energy changes in an unphysical manner depending on the relative position of the grid point and the nucleus, which prevents us from implementing the practical calculation. As shown in Fig. 4.4, the unphysical oscillation in the total energy of a copper atom is reduced substantially when the double-grid technique is used.

As the third example, the timesaving double-grid technique is applied to the calculation of the adiabatic potential curve of F_2 and Cu_2 molecules. For the calculation of the F_2 molecule, the coarse grid spacing is set at 0.33 a.u., and for the Cu_2 molecule, it is set at 0.27 a.u. Figure 4.5 shows the adiabatic potential curves for the respective molecules. In the calculation without using the double-grid technique, the total energy is mostly dependent on the relative positions of the grid points and the nuclei rather than on interatomic distances, and the equilibrium bond length cannot be determined. On the other hand, when using the double-grid technique, the curves do not depend on the relative positions, and the locations of these minima are in good agreement with the equilibrium bond length (2.68 a.u. for F_2 and 4.19 a.u. for Cu_2) obtained experimentally and by other methods. These results demonstrate that the first-principles electronic-structure calculation based on the real-space finite-difference method has been improved to be a practical tool by incorporating the timesaving double-grid technique. Moreover, in the double-grid technique, because it is not necessary to compute the Pulay force, the computational effort is very small.

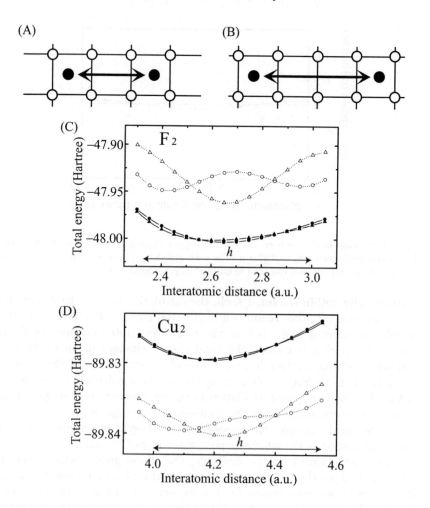

Fig. 4.5 Relative positions with respect to the grids for F_2 and Cu_2 molecules and adiabatic potential curves. (A) The center of gravity of the molecule is placed at the center of neighboring grids with respect to all the directions of x, y and z. In the figure, 'o' indicates the coarse grid and • represents nuclei. (B) The center of gravity of the molecule is placed on the grid plane which orthogonally intersects the coupling axis. For the grid plane parallel to the coupling axis, the center of gravity is placed at the center of neighboring grids. (C) Adiabatic potential curves of F_2 molecules. The grid spacing h is 0.33 a.u. (D) Adiabatic potential curve for Cu_2 molecules. The grid spacing h is 0.27 a.u. The calculated points are fit to spline functions as a guide for the eye. In (C) and (D), circles and triangles represent the results for configurations (A) and (B), respectively. The solid (dotted) curves are adiabatic potential curves calculated with (without) the double grid. Curves for Cu_2 molecules without the double grid are shifted by +0.11 a.u.

4.3   Other Mesh-Refinement Techniques

Other mesh-refinement techniques to circumvent the problem regarding unphysical oscillation of the total energy have been developed. One is a method of placing nested uniform patches of refinement locally in space. Fattebert (1999) developed a method using a composite grid and computed the equilibrium geometry of a CO molecule using a norm-conserving pseudopotential. Wang and Beck (2000) reported that surprisingly accurate results can be obtained by incorporating uniform patches into real-space all-electron calculations.

An alternative method is to concentrate grids in the vicinity of a nucleus using an adaptive curvilinear coordinate. Gygi and Galli (1995) applied this method to the calculation of the equibrium geometry and harmonic vibrational frequencies of CO_2, CO, N_2, and F_2 molecules using norm-conserving pseudopotentials and found that the results are in excellent agreement with the results of accurate quantum-chemistry and local-density-functional calculations. Modine *et al.* (1997) also demonstrated the functionality of all-electron and pseudopotential methods by calculating accurate forces, band structures, and structural properties for some standard test systems. However, in this method, the grid moves as the nucleus moves; the calculation of Pulay force, which is computationally demanding, is required.

Chapter 5

Implementation for Systems under Various Boundary Conditions

Tomoya Ono and Kikuji Hirose

5.1 Isolated Boundary Condition: Molecules

We here demonstrate the viability of the real-space finite-difference method and assess the sort of accuracy obtainable. Our first series of tests is performed on isolated diatomic molecule systems under isolated boundary conditions. The formation of the diatomic molecule is accompanied by a decrease of total energy. The properties of these systems are often calculated using the basis set of atomic orbitals in quantum chemistry. The results obtained using the real-space finite-difference method are compared with the experimental result and the other theoretical results obtained using basis sets.

The computational conditions are as follows: the coarse-grid spacing for N_2 and F_2 molecules is set at 0.33 a.u. and those for the other molecules are set at 0.35 a.u. The dense-grid spacing is fixed at $h_{dens} = h/3$, where h denotes the coarse-grid spacing. We obey the nine-point finite-difference formula, i.e., $N_f = 4$ in (1.5), for the derivative arising from the kinetic-energy operator. The norm-conserving pseudopotential [Kobayashi (1999a)] of Troullier and Martins (1991) is employed in a separable form of Kleinman and Bylander [see Eq. (1.23)]. Exchange-correlation effects are treated with the local-spin-density approximation [Perdew and Zunger (1981)].

Table 5.1 lists the calculated results of bond lengths, vibrational modes, and cohesive energies for diatomic molecules consisting of first-row elements. Since the local-spin-density approximation overestimates cohesive energies, the results of cohesive energies are not in good agreement with those of experiments, but are consistent with the results of the other first-

Table 5.1   Properties of diatomic molecules. The experimental data are from Huber and Herzberg (1979). The theoretical results are from our present calculation and from other methods using similar forms of the local-spin-density approximation.

	N_2	O_2	CO	F_2
Bond length (a.u.)				
Experiment	2.07	2.28	2.13	2.68
This work	2.07	2.28	2.14	2.66
Other theory	2.07[‡]	2.27[‡]	2.13[†]	2.64[†]
Vibrational mode (cm^{-1})				
Experiment	2358	1580	2169	892
This work	2400	1626	2160	1127
Other theory	2380[‡]	1620[‡]	2151[†]	1051[†]
Cohesive energy (eV)				
Experiment	9.76	5.12	11.09	1.60
This work	11.32	7.44	12.93	2.66
Other theory	11.6[‡]	7.6[‡]	12.8[‡]	3.4[‡]

[†]From Gygi and Galli (1995).
[‡]From Becke (1986, 1992).

Table 5.2   Properties of diatomic molecules. The experimental data are from Huber and Herzberg (1979).

	Cu_2	CuH	$CuCl_2$
Bond length (a.u.)			
Experiment	4.19	2.77	3.88
This work	4.19	2.77	3.87
Vibrational mode (cm^{-1})			
Experiment	266	1940	418
This work	303	1958	431
Cohesive energy (eV)			
Experiment	1.97	2.73	3.82
This work	2.69	3.28	4.14

principles calculations. The other properties are in accord with experimental values [Huber and Herzberg (1979)] and the results obtained using other calculations.

We next examine the properties of diatomic molecules including transition metals. Since transition metals have d valence electrons, it is important to examine the viability for a system including these metals. We here take the coarse-grid spacing of 0.27 a.u. Table 5.2 shows the calculated results, which are in agreement with the experimental results except for the cohesive energies.

All these results demonstrate the applicability of the real-space finite-difference method for studying the geometrical and electrical properties of isolated molecule or cluster systems. For studies using isolated boundary conditions, see Bernholc *et al.* (1991), Chelikowsky *et al.* (1994a, 1994b, 1996), Gygi and Galli (1995), Seitsonen *et al.* (1995), Modine *et al.* (1997), Jin *et al.* (1999), Takahashi *et al.* (2001, 2002, 2003), and Hori *et al.* (2003a, 2003b).

5.2   3D Periodic Boundary Condition: Crystals

Our second series of tests is the calculation of the bulk properties of Na, Al, Si, Fe, Co, and Ni crystals under the three-dimensional periodic boundary conditions. These systems are usually treated with the plane-wave basis set. In order to ensure the efficiency of the real-space finite-difference approach, the computed bulk properties are compared with the experimental and other theoretical results.

The computational conditions are as follows: the coarse-grid spacing h is set at ~ 0.30 a.u., and the dense-grid spacing is fixed at $h_{dens} = h/3$. The k-space integrations are performed with 55 **k**, 89 **k**, and 139 **k** points in the irreducible Brillouin zone for the body-centered-cubic (bcc), face-centered-cubic (fcc), and diamond structures, respectively. The nine-point finite-difference formula and the local-spin-density approximation [Perdew and Zunger (1981)] are employed. The norm-conserving pseudopotential [Kobayashi (1999a)] of Troullier and Martins (1991) is employed in a separable form of Kleinman and Bylander [see Eq. (1.23)]. Coulomb potentials and energies in infinite systems are computed using the formulas given in Appendix A.1.

The results are given in Tables 5.3 and 5.4 together with results from experiments and other calculations. Figures 5.1 and 5.2 show the electronic band structures. These results are consistent with experimental values and those obtained using other calculations. In addition, for all studied metals, we note good agreement between the band structures obtained by the real-space finite-difference method and by the plane-wave expansion method [see, for example, Moroni *et al.* (1997)].

For studies using the three-dimensional periodic boundary conditions, see Briggs *et al.* (1995, 1996), Modine *et al.* (1997), Jin *et al.* (1999), and Ono *et al.* (2003a).

Table 5.3   Calculated lattice constants, bulk moduli, and cohesive energies for bcc Na, fcc Al and diamond-structure Si.

	Na	Al	Si
Lattice constant (a.u.)			
Experiment	7.98[†]	7.65[†]	10.26[†]
This work	7.70	7.63	10.19
Other theory	7.69[‡]	7.58[¶]	10.19[§]
Bulk modulus (Mbar)			
Experiment	0.07[†]	0.72[†]	0.99[†]
This work	0.09	0.74	0.96
Other theory	0.09[‡]	0.72[¶]	0.98[§]
Cohesive energy (eV/atom)			
Experiment	1.11[†]	3.39[†]	4.63[†]
This work	1.25	3.90	5.32
Other theory	1.21[‡]	3.65[¶]	5.9[§]

[†] From Kittel (1986).
[‡] From Sigalas *et al.* (1990).
[¶] From Lam and Cohen (1981).
[§] From Holzwarth *et al.* (1997).

Table 5.4   Calculated lattice constants, bulk moduli, magnetic moments, and cohesive energies for ferromagnetic bcc Fe, fcc Co and Ni.

	Fe	Co	Ni
Lattice constant (a.u.)			
Experiment	5.42[†]	6.69[†]	6.65[†]
This work	5.20	6.62	6.58
Other theory	5.22[§]	6.52[§]	6.48[§]
Bulk modulus (Mbar)			
Experiment	1.68[†]	1.87[‡]	1.86[†]
This work	2.68	2.50	2.70
Other theory	2.35[§]	2.42[§]	2.55[§]
Magnetic moment (μ_B)			
Experiment	2.22[†]	1.75[¶]	0.61[†]
This work	2.07	1.61	0.54
Other theory	2.05[§]	1.52[§]	0.59[§]
Cohesive energy (eV/atom)			
Experiment	4.28[†]	—	4.44[†]
This work	6.06	6.47	6.86
Other theory	6.47[§]	—	6.09[§]

[†] From Kittel (1986).
[‡] From Guillermet and Grimvall (1989).
[¶] From Stearns (1986).
[§] From Moroni *et al.* (1997).

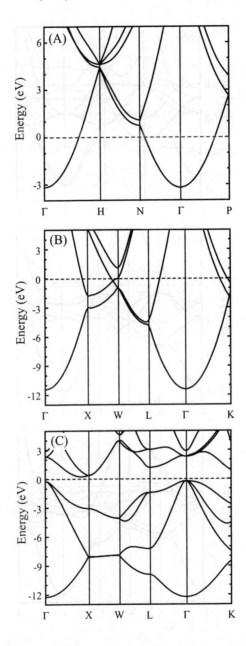

Fig. 5.1 Band structures for (A) bcc Na, (B) fcc Al, and (C) diamond Si with experimental lattice parameters. The zero of energy is chosen to be the Fermi level E_F.

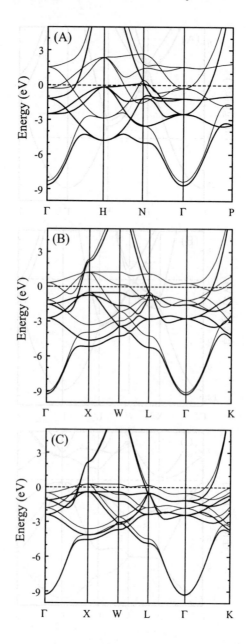

Fig. 5.2   Band structures for ferromagnetic (A) bcc Fe, (B) fcc Co, and (C) fcc Ni with experimental lattice parameters. The thick curves are the majority spin bands while thin curves are for minority spin. The zero of energy is chosen to be the Fermi level E_F.

5.3   (2D Periodic + 1D Isolated) Boundary Condition: Thin Films

In the case of simulations under external electric fields using conventional plane-wave basis sets, the periodic boundary condition gives rise to saw-tooth potential, which sometimes leads to numerical instability during the self-consistent iteration. On the other hand, the real-space finite-difference method can exactly determine the potential by the external electric field as a boundary condition and is free from involving the saw-tooth potential. The applicability to simulations under electric fields is one of the advantages of the real-space finite-difference method. In this section, a simulation for the evaporation of surface atoms from Si(001) surfaces, in which the surfaces are imitated with thin films, is implemented as an application for the two-dimensional periodic boundary condition.

So far, scanning tunneling microscopy experiments in the atomic-scale manipulation of material surfaces have attracted great attention due to their potential to create artificial surface nanostructures. Eigler and Schweizer (1990), and Lyo and Avouris (1991) demonstrated that it is possible to remove adsorbed atoms and molecules from surfaces or to deposit them on surfaces by applying voltage pulses. Field evaporation [Tsong (1990)] and deposition are thermally activated processes in which rate constants can be parameterized according to the Arrhenius formula [Arrhenius (1889)]. The electric field between probing tip and surface is considered to play an important role during these processes, since the activation energy for field evaporation is considered to depend on this electric field. Kawai *et al.* [Watanabe and Satoh (1993); Kawai and Watanabe (1996); Kawai and Watanabe (1997); Kawai *et al.* (1998)] examined the adiabatic potential curves for adsorbed atoms on the atomically flat Si(001) surface, however, the maximum external electric field of ~ 2.6 V/Å applied in their simulation is rather small compared to that in experiments using field ion microscopy (3.8 V/Å) [Tsong (1990)].

Here we study activation energies for silicon atoms on Si(001) surfaces to evaporate under the stronger electric fields of 3.0, 3.5, and 4.0 V/Å. Moreover, we explore differences in activation energies and threshold values of external electric fields among silicon atoms evaporating from various Si(001) surfaces, e.g., the atomically flat surface, the surface having a step, and the atomically flat surface with an adsorbed silicon atom.

The computational conditions are as follows: the nine-point finite-difference formula for the derivative arising from the kinetic-energy operator is adopted. We take a grid spacing of 0.65 a.u., and a denser grid spacing of 0.22 a.u. in the vicinity of nuclei with the augmentation of double-grid points. The norm-conserving pseudopotential [Kobayashi (1999a)] of

Troullier and Martins (1991) is employed in a separable form of Kleinman and Bylander [see Eq. (1.23)]. Coulomb potentials and energies in infinite systems are computed using the formulas given in Appendix A.2. The local density approximation [Perdew and Zunger (1981)] is employed to treat exchange-correlation effects. Figure 5.3 shows the top views of the Si(001) surfaces calculated here. We employ a technique that involves the use of a supercell whose size is chosen as $L_x = 43.52$ a.u., $L_y = 14.51$ a.u. and $L_z = 32.64$ a.u., where L_x, L_y and L_z are the lengths of the supercell in the x, y and z directions, respectively. Here the direction perpendicular to the surface is chosen as the z direction. In order to eliminate completely unfavorable effects of atoms in neighbor cells which are artificially repeated in the case of the periodic boundary condition, the isolated boundary condition of vanishing wave functions out of the supercell is imposed in the z direction, while the periodic boundary condition in the x and y directions is employed. The supercell contains five silicon layers, the lowest of which is terminated by the hydrogen atom, i.e., the thin film model. The forces acting on atoms in the three topmost silicon layers are fully relieved by the first-principles structural optimization.

First an external electric field F of 3.0 V/Å is applied along the z direction and it is found that no surface atoms evaporate in any models. In the case of $F = 3.5$ V/Å, the adsorbed atom in the adatom model evaporates [see Fig. 5.4(A)]. From these results, the threshold value of the external electric field for evaporation is 3.0 – 3.5 V/Å. This value agrees with the experimental data of 3.8 V/Å obtained using a field ion microscopy. When the external electric field F is increased to 4.0 V/Å, the two upper atoms of the buckled dimers located at the step edges simultaneously evaporate in the S_A step model [see Fig. 5.4(B)]. Differences in activation energies for the field evaporation between them are negligibly small, although each of two upper atoms of the dimers is located in a slightly different situation. It is evident that the surface atoms in the S_A step model are preferentially removed from the step edges. The evaporated atoms in both the adatom and S_A step models are approximately doubly charged. In the terrace model, surface atoms do not evaporate at all below $F = 4.0$ V/Å.

It is noted that one of the potentially important applications of the field-evaporation process is the direct determination of the binding strengths of the surface atoms from the external electric field required for their removal. We now evaluate the activation energies of surface atoms for field evaporation during the lifting up of the surface atoms. The activation energies for various surface models are shown in Table 5.5. The activation energies in the absence of the external electric field are equal to the binding energies between evaporating atom and surface. They are observed to be 6.03, 5.86, and 5.04 eV for the terrace, S_A step, and adatom model, re-

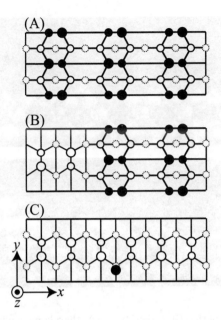

Fig. 5.3 Top views of the three topmost layers of Si(001). (A) The atomically flat surface (terrace model), (B) the surface having a S_A step (S_A step model), and (C) the surfaces where an atom is adsorbed (adatom models). Closed circles, opened circles with solid curves, and opened circles with dotted curves represent atoms on the top, second, and third layers, respectively. Large (small) circles represent the upper (lower) atoms in buckled dimers.

spectively, and these energies become lower as the external electric fields are increased. The threshold value of the external electric field varies with the bond strength of the surface atom. According to the Arrhenius formula [Arrhenius (1889)], the rate constants of field evaporation depend on the activation energies; surface atoms in both the S_A step model and the adatom model are easily removed compared with the surface atom in the terrace model. It is noteworthy that our simulations are implemented at zero Kelvin, and surface atoms evaporate with low electric field at high temperatures since the temperature has an effect on the field evaporation process [Sudoh and Iwasaki (2000)]. From what has been mentioned above, we conclude that the field-evaporation process can be applied to surface-flattening techniques or atomic-scale manipulation methods by adjusting the strength of the external electric field and the temperature.

We show in Fig. 5.5 the electron density shift $\rho(\mathbf{r}, F) - \rho(\mathbf{r}, F = 0)$ due to the application of the external electric field of $F = 3.5$ V/Å. The overall charge around the surface atoms decreases and the atom is expected

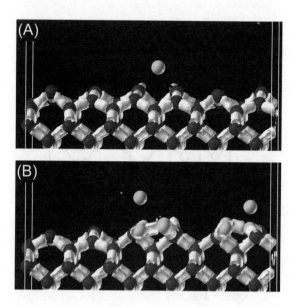

Fig. 5.4   Isosurfaces of the charge density for (A) the adatom model under $F = 3.5$ V/Å and (B) the S_A step model under $F = 4.0$ V/Å. Light and dark balls represent the surface atom and the atoms in the silicon surface, respectively. Reprinted with permission from Ono and Hirose (2004a). © 2004 American Institute of Physics.

Table 5.5   Activation energies (in eV). The values at zero field are the binding energies between the lifted-up atoms and the surfaces. For the S_A step model, we depict the activation energies of the center atom in Fig. 5.3(B). Data taken from Ono and Hirose (2004a).

	0.0(V/Å)	3.0(V/Å)	3.5(V/Å)	4.0(V/Å)
Terrace	6.03	1.45	0.52	0.24
S_A step	5.86	1.05	0.13	0.00
Adatom	5.04	0.09	0.00	0.00

to be a positive ion when it evaporates. We then calculate the external electrostatic field as the difference between the total electrostatic field in the presence of an external electric field and that in the absence of it, $d(V_{eff}(\mathbf{r}, F) - V_{eff}(\mathbf{r}, F = 0))/dz$, where V_{eff} is the sum of the external, Hartree, and exchange-correlation potential functions. Figure 5.6 shows the counter plots of the difference in external electrostatic field. One can clearly recognize in all models the expulsion of the external electronic field from inside the surface. However the local-field enhancement occurs not around the evaporating atom but above it, and no further significant differ-

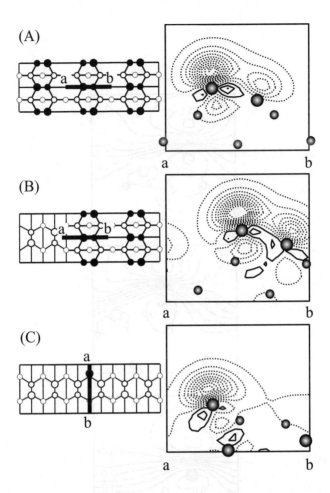

Fig. 5.5   Electronic density shift $\rho(\mathbf{r}, F) - \rho(\mathbf{r}, F = 0)$ at $F = 3.5$ V/Å   represented on the cross section of the ribbon containing the thick line a–b. The contour spacing is 6.7 electron/supercell. Solid (dotted) curves represent nonnegative (negative) values. The large and small balls indicate the atomic positions on and above the cross section, respectively. Reprinted with permission from Ono and Hirose (2004a). © 2004 American Institute of Physics.

ence in external electrostatic field around the evaporating atom is observed. Thus the local field around the evaporating atom does not play a crucial role in the situation mentioned above. These results present the potential power of the real-space finite-difference method for the simulations under the external electric field. For further details of this simulation, see Ono and Hirose (2004a).

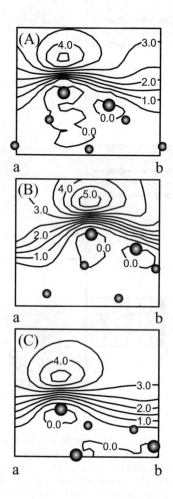

Fig. 5.6   Difference in external electrostatic field between $F = 0.0$ V/Å and $F = 3.5$ V/Å. The contour spacing is 0.5 V/Å. The meanings of the symbols are the same as those in Fig. 5.5. Reprinted with permission from Ono and Hirose (2004a). © 2004 American Institute of Physics.

5.4 (1D Periodic + 2D Isolated) Boundary Condition: Infinite Metallic Wires

In the last several years, the properties of metallic wire contacts have attracted great attention, and many experiments concerning wire contacts have been carried out using a scanning tunneling microscopy and a mechanically controllable break junction [see, for example, van Ruitenbeek (2000) and references therein]. The simultaneous measurement of mechanical force and conductance and also the direct observation by *in situ* transmission electron microscopy gave us visible information about the relationship between geometrical structure and electrical conduction of the wire. In particular, an infinite single-row atomic wire (ISAW) has attracted great attention from a fundamental point of view, because phenomena peculiar to the reduced dimensionality are expected to emerge best for the ISAW having an ultimately simplified one-dimensional structure [Ichimura *et al.* (1998); Sánchez-Portal *et al.* (1999); Torres *et al.* (1999); Maria and Springborg (2000); Okada and Oshiyama (2000); Watanabe *et al.* (2000); Sen *et al.* (2001); Tsukamoto *et al.* (2001); Ayuela *et al.* (2002); Ono *et al.* (2003b); Okano *et al.* (2004)]. So far, a large number of theoretical studies concerning the infinite wires have been reported and it is well known that the wires made of monovalent atoms distort to form dimer, which is so called Peierls distortion [Fröhlich (1954); Peierls (1955)]. On the other hand, the characteristic of the aluminum with $3s$- and $3p$-valence electrons is expected to be different from that of monovalent atoms. Therefore, an ISAW consisting of aluminum atoms (Al-ISAW) is an important element for understanding distortion of the ultimate one-dimensional (1D) wires. We present here a theoretical analysis of the interesting relationship between geometrical and spin-electronic structures of Al-ISAW as an application to 1D periodic boundary condition.

The computational conditions are as follows: the norm-conserving pseudopotential [Kobayashi (1999a)] of Troullier and Martins (1991) is employed in a separable form of Kleinman and Bylander [see Eq. (1.23)]. Coulomb potentials and energies in infinite systems are computed using the formulas given in Appendix A.3. We obey the nine-point finite-difference formula for the derivative arising from the kinetic-energy operator. We take a grid spacing of 0.6 a.u., and a denser grid spacing of 0.2 a.u. in the vicinity of the nuclei with the augmentation of double-grid points. Exchange-correlation effects are treated with the generalized-gradient approximation [Perdew *et al.* (1992)] in which spin degrees of freedom are taken into account. We employ a technique that involves the use of a supercell whose size is chosen as $L_x = L_y = 20$ a.u. and $L_z = N_{atom} \times d_{av}$. Here, the z axis is taken along the wire axis, the x, y axes are perpendicular to the wire, L_x, L_y and

L_z are the lengths of the supercell in the x, y and z directions, respectively, N_{atom} is the number of aluminum atoms within the supercell, and d_{av} is the average value for *projections* of the interatomic distances between adjacent atoms onto the z component. We impose the isolated boundary condition of vanishing wave function out of the supercell in the x and y directions to eliminate completely unfavorable effects of atoms in neighbor cells which are artificially repeated in the case of periodic boundary condition, while we adopt the periodic boundary condition in the z direction. We implement structural optimization until the remaining forces acting on atoms are smaller than 16.5 pN. The Brillouin-zone integration is performed using equidistant k_z-point sampling where the k_z-point density is equivalent to the case of the 60-point sampling in the irreducible Brillouin zone of the regular wire. The convergence with respect to the number of k_z points was assured.

In order to gain an insight into the Peierls distortion of the Al-ISAW, the total energies of the wires with several atomic and spin-electronic structures are examined using large supercells of $N_{atom} \geq 2$. While the atoms in the wire were allowed only to dimerize in the previous studies by Sen *et al.* (2001) and Ayuela *et al.* (2002), we consider additional geometrical degrees of freedom, such as trimerization and tetramerization, for definite clarification of the atomic and spin-electronic structures of the Al-ISAW during its elongation. Taraschi *et al.* (1998) also studied structural properties of the Al-ISAW of $N_{atom} = 5$ at $d_{av} = 4.55$ a.u., however, the possibility of spin polarization was not considered and the average interatomic distance was not long enough for the Al-ISAW to rupture in their study. We here take into account spin degrees of freedom and choose an average interatomic distance to be sufficiently large. Figure 5.7 depicts our calculation models and results. We take the average interatomic distance d_{av} of 5.6 a.u., which is longer than the critical distance of ~ 5.4 a.u. where an ISAW transforms from a zigzag structure into a linear one [Ono and Hirose (2003)]. In Fig. 5.7(B), the middle atoms within the trimers are located at the centers of the respective trimers in an intertrimer distance of $3d_{av}$. In the case of the tetramized ISAW in Fig. 5.7(C), the two middle atoms inside the tetramers are relaxed during the structural optimization, whereas both the edge atoms of the tetramers are frozen so as to maintain a tetramer length of $3(d_{av} - \Delta d)$ and an intertetramer distance of $4d_{av}$. We observe the electronic ground states with paramagnetic (PM), ferromagnetic (FM) and antiferromagnetic (AFM) orderings, but do not find that with the ferrimagnetic ordering. There is a sensible difference in energy between regular and distorted wires: the total energy per atom of the trimerized wire with AFM ordering is the lowest as shown in Fig. 5.7(D). In addition, we ensured that the trimerized wire with AFM ordering is also the most stable structure in the range of 5.3 a.u. $< d_{av} < 5.8$ a.u.

Let us consider why the Al-ISAW prefers a trimerized structure with AFM ordering after it breaks. This reason is understood by noting the behavior of the σ-symmetry bands. In Fig. 5.8(D), one can recognize that the regular wire has doubly-degenerate π-symmetry bands and nondegenerate σ-symmetry ones near the Fermi level E_F, and the π-symmetry and σ-symmetry bands cross the E_F near $k_z \approx \frac{\pi}{3d_{av}}$ and $k_z \approx \frac{\pi}{6d_{av}}$, respectively. When the ISAW ruptures, the spin-density wave (SDW) of $6d_{av}$ periodicity appears so as to open up an energy gap of σ-symmetry bands at the E_F and to lower the total energy. We show in Fig. 5.8(B) the energy band structure of the trimerized ISAW with AFM ordering at $d_{av} = 5.6$ a.u. One can find the energy band gap at the E_F. On the other hand, the SDWs which have the other periodicity, such as $2d_{av}$ and $3d_{av}$, lead the σ-symmetry bands to cross the E_F (see Fig. 5.8) and increase the total energy of the ISAW relative to the trimerized ISAW with AFM ordering having SDW of $6d_{av}$ periodicity. We conclude that the preference for a trimerized structure with AFM ordering in the distorted wire is attributed to the nature of its σ-symmetry bands. For further details of this simulation, see Ono and Hirose (2003).

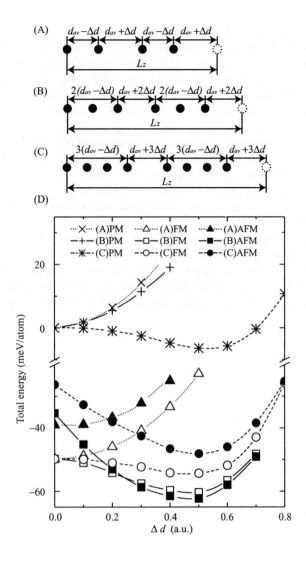

Fig. 5.7 Models for (A) dimerized wire of $N_{atom} = 4$, (B) trimerized wire of $N_{atom} = 6$, and (C) tetramized wire of $N_{atom} = 8$. (D) Total energy per atom as a function of the bond length of the dimer in the wire with PM, FM, and AFM orderings. The zero of energy is taken to be the total energy of the wire with PM ordering at $\Delta d = 0$ a.u. The average interatomic distance d_{av} is set to be 5.6 a.u. In (A), (B), and (C), the closed circles denote the atoms in a supercell and the open one shows the replicated atom in the adjacent supercell. Reprinted with permission from Ono and Hirose (2003). © 2003 American Physical Society.

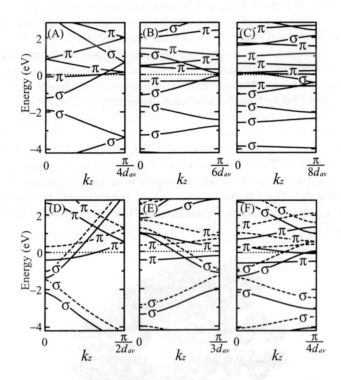

Fig. 5.8   Energy band structures of the Al-ISAW: (A) dimerized wire at $\Delta d = 0.0$ a.u. with AFM ordering, (B) trimerized wire at $\Delta d = 0.5$ a.u. with AFM ordering, (C) tetramized wire at $\Delta d = 0.5$ a.u. with AFM ordering, (D) dimerized wire at $\Delta d = 0.0$ a.u. with FM ordering, (E) trimerized wire at $\Delta d = 0.5$ a.u. with FM ordering, and (F) tetramized wire at $\Delta d = 0.4$ a.u. with FM ordering plotted along the direction k_z in reciprocal space. Same as Fig. 5.7 for the definition of Δd. In (D), (E), and (F), solid and dashed curves represent up-spin and down-spin electron bands, respectively. The zero of energy is taken to the Fermi level E_F. The average interatomic distance d_{av} is set to be 5.6 a.u. Reprinted with permission from Ono and Hirose (2003). © 2003 American Physical Society.

5.5   Twist Boundary Condition: Infinite Helical Carbon Nanotubes

Carbon nanotubes consist of a few concentric tubes each of which has carbon-atom hexagons arranged in a helical fashion about an axis [Iijima (1991)]. Their topologies are characterized by a set of two numbers (n, m) called chiral indices. Since the moment they were first discovered, their atomic and electronic structures have been intensively explored by both experimental and theoretical approaches. Previous theoretical studies have shown that carbon nanotubes change their nature between metallic and semiconducting depending on their structure, such as the diameter and the helical arrangement [Hamada *et al.* (1992); Saito *et al.* (1992)]. We here examine the electronic band structure of an infinitely long $(6, 6)$ carbon nanotube, which is depicted in Fig. 5.9, by the real-space finite-difference approach utilizing the twist boundary condition. Such an approach is particularly appropriate for the helical nanotubes, because, although the nanotubes can be constructed with 1D translational symmetry, the translational repeat distance leads to extremely large supercell sizes.

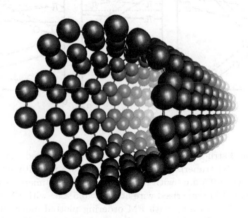

Fig. 5.9   Structure of $(6, 6)$ carbon nanotube.

The computational conditions are as follows: the norm-conserving pseudopotential [Kobayashi (1999a)] of Troullier and Martins (1991) is employed in a separable form of Kleinman and Bylander [see Eq. (1.23)]. Coulomb potentials and energies in infinite systems are computed using the formulas given in Appendix A.4. The tube structure is generated by a screw operation with a twist of $\pi/6$ rad and a translational shift of 2.32 a.u., which corresponds to the length of the supercell L_z in this case and which was chosen such that nearest-neighbor separations between rings are equal to the in-ring values. Exchange-correlation effects are treated by the local density approximation [Perdew and Zunger (1981)] and the central finite-difference formula, i.e., $N_f = 1$ in (1.5), is adopted. We chose a coarse grid spacing of 0.33 a.u. and a denser grid spacing of 0.11 a.u. in the vicinity of the nuclei with the augmentation of double-grid points.

The helical band structure is depicted in Fig. 5.10(A). Since there are multiple equivalent ways of depicting the band structure of the (n, n) carbon nanotube, Mintmire *et al.* [Mintmire *et al.* (1992); Mintmire and White (1998)] proposed the 'pseudo' band structure. The procedure for drawing a pseudoband structure is as follows: the electronic band structure is computed for a helical structure generated with a helical twist angle of φ, using the irreducible representation of the C_n rotation group. This band structure will also be effective for the helical band structure generated with a different angle of $\varphi + \Delta\varphi$, where $\Delta\varphi = 2\pi l/n$ and l is a integral between 0 and $n-1$, but some states will occur at a shifted wave vector $k' = k + j\Delta\varphi$. One can classify all the bands into two representations: one is the a representation which is invariant, in the Brillouin zone, to the choice of the helical twist angle and the other is the e representation which moves in the Brillouin zone depending on the choice of the angle. The pseudoband structure is obtained by shifting all the e bands in k by a quantity corresponding to a phase factor of $\pm j\varphi$. This choice of phase shifts the e bands to doubly degenerate bands.

The pseudoband structure of the $(6, 6)$ carbon nanotube is illustrated in Fig. 5.10(B). We also depict, in Fig. 5.11, the band structure obtained using a conventional supercell which has twice the length, $2L_z$, in the tube direction with a helical twist angle of zero. One can recognize that the pseudoband structure leads to a depiction of the bands as equivalent as possible to the bands that would be present if the system were to have a translational periodicity with a repeat distance of the helical supercell.

This work demonstrates the utility of the real-space finite-difference method in obtaining the ground-state electronic structure of carbon nanotubes possessing a helical twist angle. When combined with conductance calculations, this approach should be useful for studying the electron transport behavior in helical nanotubes.

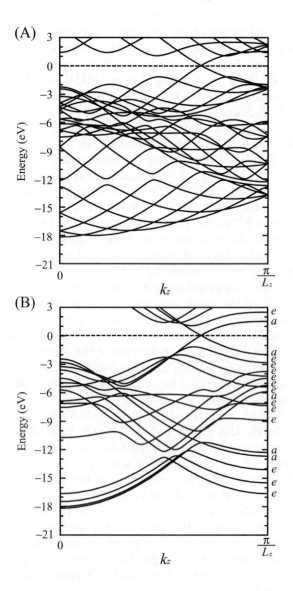

Fig. 5.10　(A) Electronic band structure of $(6,6)$ carbon nanotube. (B) Pseudoband structure of $(6,6)$ carbon nanotube. The zero of energy is chosen to be the Fermi level E_F.

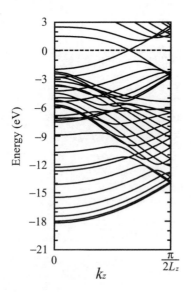

Fig. 5.11 Electronic band structure of $(6, 6)$ carbon nanotube using the conventional supercell with the length of $2L_z$. The zero of energy is chosen to be the Fermi level E_F.

PART II

Electronic Transport through Nanostructures between Semi-Infinite Electrodes

Chapter 6

Basic Scheme of the Overbridging Boundary-Matching Method

Yoshitaka Fujimoto, Tomoya Ono and Kikuji Hirose

6.1 Preliminary Concepts: Ballistic Transport of Electrons and the Landauer Formula

Let us first consider the mechanism of electron conduction in a macroscopic system. In solids, such as metals and semiconductors, electrons in the vicinity of the Fermi level are accelerated by an applied electric field, and undergo inelastic scattering with energy dissipation caused by lattice vibration (phonon) and collisions with impurities. While repeating the process in which electrons are accelerated and inelastically scattered, electrons in general proceed in the direction of the electric field (drift conduction) and reach drift speed, which is proportional to the magnitude of the electric field. As a result, for a conductor wire that is longer than the mean free path of electrons (which is the average distance that electrons proceed without being subjected to scattering; 10 – 50 nm for metallic bulks at room temperature), Ohm's law holds; electric conductance is inversely proportional to the length of the wire and proportional to its cross-sectional area.

In contrast, for nanostructures, in which fine nanowires shorter than the mean free path are connected to electrodes, most of electrons proceed without inelastic scattering; instead, they penetrate the nanowires ballistically from one end to the other. This type of conduction is called ballistic transport of electrons. In the case of the ballistic transport, the conductance is independent of the length of the nanowires. Furthermore, when the diameter of the cross section of the nanowires becomes as small as the Fermi wavelength of electrons (for example, the Fermi wavelength for gold is 0.52 nm) and thus energy levels are quantized, electrons can pass through the nanowires only via the quantized energy levels. Therefore, a conduction phenomenon that is significantly different from that having the general macroscopic diffusive-transport characteristics of solids, should be observed in ballistic transport. Actually, coherence of electron movement, which is not observed in macroscopic systems, is conspicuous, and several interesting phenomena, such as quantization of the conductance, unusual nonlinear current–voltage characteristics and Coulomb blockade effect, arise.

The experimental verification of the quantized conductance was first given by van Wees *et al.* (1988) in the two-dimensional electron gas system using semiconductor crystals. Since the mean free path of electrons in a semiconductor is rather long, the formation of a narrow path through which electrons can pass is relatively easy. For metals, the mean free path is as small as 10 – 50 nm; fine nanowires must be formed. Pascual *et al.* (1993) developed an ingenious method, in which a scanning tunneling microscopy (STM) metal probe is pulled up after coming into contact with the surface of the metal substrate, and observed quantization of the conductance for gold nanowires thus formed. Here, the conductance changes stepwise just prior to the breakage of the metal nanowire. Furthermore, the height of one step is G_0 $(= 2e^2/h)$, demonstrating that the conductance is quantized, where e is the electron charge and h is Planck's constant. Subsequently, using a STM and a mechanically controllable break junction, similar observations for various metals were achieved; in many of the metals, quantization of conductance was confirmed. For reviews, see, for example, van Ruitenbeek (2000) and Agraït *et al.* (2003).

Theories developed by Landauer and Kubo make up the basic theory of electronic transport through metals and semiconductors which is widely used today [Landauer (1957); Kubo (1957)]. The contents of the two theories differ in that; Landauer's theory provides intuitive and simple formulas, while Kubo's theory is a universal theory in which involved mathematical techniques are required. The former is powerful for low-dimensional problems, particularly one-dimensional problems. In what follows, we briefly discuss Landauer's theory. For reviews, see Datta (1995), Imry (1997), Kawabata (2000) and Agraït *et al.* (2003).

Figure 6.1 illustrates Landauer's model. The major feature of Landauer's theory is the introduction of electrodes at both ends of a one-dimensional conductor for dc conduction, where the conductor is composed of a scatterer and left and right leads. This setup is in contrast to Kubo's model in which an infinitely long conductor is assumed. In actual experiment, since batteries or power sources are connected to a sample of a finite size via lead wires, Landauer's model is closer to the real situation. What drives the current is the difference of chemical potentials, μ_L and μ_R, of the left and right electrodes, respectively. We assume that the electrodes are *ideal* electron reservoirs having the following characteristics:

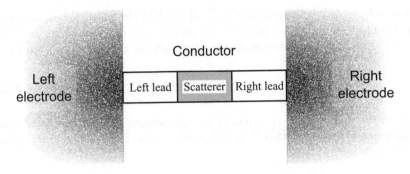

Fig. 6.1 Schematic view of Landauer's model. The scatterer is connected with two electrodes by the leads.

(i) Any electron with energy $E < \mu_L$ (μ_R) is supplied to the left (right) lead from the left (right) electrode.

(ii) All of electrons entering the electrodes from the leads are accepted by the electrodes.

(iii) Inside the electrodes, inelastic scattering highly frequently occurs and thermal equilibrium is achieved rapidly. The capacity of the electron reservoirs is so large that the chemical potentials of the electrodes are maintained at μ_L and μ_R, independently of the input and output of electrons through the leads.

We further assume that the leads are also *ideal* ones such that electrons are not scattered inside the leads, but merely pass through the leads; in the scatterer, only elastic scattering can take place and hence electrons move coherently between the electrodes. For simplicity, the conductor is assumed to have a single channel, in other words, the number of modes in the directions perpendicular to the conductor (transverse modes) is one.

The elastic scattering process of electrons in the scatterer is considered to be expressed by the transmission probability, $\mathcal{T}$, and reflection probability $\mathcal{R}$ ($\mathcal{T} + \mathcal{R} = 1$). Supposing $\mu_L > \mu_R$, the flow of electrons with energy $E < \mu_R$ is completely symmetrical and the net flow is zero. For electrons with $\mu_R < E < \mu_L$, only electrons headed to the right are supplied by the left electrode. Among them, some electrons are transmitted with probability $\mathcal{T}$ and enter the right electrode. Let us denote the wave number and energy of such electrons by k and $E(k)$, respectively. Thus the magnitude of the current carried by them is

$$I(k) = -e\mathcal{T}\frac{dE(k)}{\hbar dk}. \tag{6.1}$$

Here, $dE(k)/\hbar dk$ is the group velocity of incident electrons. When the wave numbers at energies equivalent to μ_L and μ_R are denoted by k_L and k_R, respectively, the total current flowing through the conductor is given by

$$I_{total} = 2 \int_{k_R}^{k_L} I(k) \frac{dk}{2\pi}, \qquad (6.2)$$

where the factor 2 on the right-hand side stands for the degree of spin degeneracy. Assuming that $\mathcal{T}$ is independent of k, from (6.1) and (6.2) one can easily see

$$I_{total} = -\frac{2e}{h} \mathcal{T} \int_{\mu_R}^{\mu_L} dE(k) = -\frac{2e}{h} \mathcal{T}(\mu_L - \mu_R). \qquad (6.3)$$

Since the potential difference between the two electrodes is given by

$$V = \frac{\mu_L - \mu_R}{-e}, \qquad (6.4)$$

the following expression for the conductance is derived.

$$G = \frac{I_{total}}{V} = \frac{2e^2}{h} \mathcal{T}. \qquad (6.5)$$

Equation (6.5) is the well-known Landauer–Büttiker formula. For more detailed studies on the Landauer–Büttiker formula, see Landauer (1970), Fisher and Lee (1981), Büttiker *et al.* (1985) and McLennan *et al.* (1991). The Landauer–Büttiker formula (6.5) implies that the conductance is quantized to G_0 in the case of $\mathcal{T} = 1$, in which there arises no elastic scattering inside the one-dimensional conductor. Note that when $\mu_L - \mu_R$ is extremely small, $\mathcal{T}$ in (6.5) is the value at the Fermi level. For metal nanowires, much evidence of the conductance quantization have been found; however, depending on the kind of metal, the value of quantization in a single channel deviates from G_0, which indicates that $\mathcal{T}$ is not always 1. In the following sections, we will give a procedure for calculating $\mathcal{T}$ for realistic models of nanostructures intervening between two crystalline electrodes.

Formula (6.5) for the conductance of a single channel is extended to the case of multichannels; postulating the transmission coefficient t_{ij} to be the probability amplitude at which electrons are transmitted from an initial mode j to a final mode i inside the conductor, one finds that the conductance is expressed by

$$G = \frac{2e^2}{h} \sum_{i,j} |t_{ij}|^2 \frac{v_i}{v_j}, \qquad (6.6)$$

where v_j and v_i are the longitudinal components of the group velocity (or the flux through the cross section of the conductor) of the modes j and i, respectively. The last factor in (6.6) accounts for the flux normalization at each channel. Equation (6.6) is called the Landauer–Büttiker formula [Büttiker *et al.* (1985)].

6.2 Wave-Function Matching Procedure

The aim of this chapter is to present, in a fairly complete fashion, details of the overbridging boundary-matching (OBM) method, which is an efficient and highly accurate first-principles treatment of ballistic transport of electrons through nanostructures sandwiched between two electrodes [Fujimoto and Hirose (2003a, 2003b); Ono and Hirose (2004b)]. The method is formulated within the framework of the density functional theory, based on the real-space finite-difference approach discussed in Part I. By virtue of a great simplicity of this approach, the wave-function matching procedure in the OBM method can be readily described and its implementation can be easily performed. The formulation includes the norm-conserving pseudopotential techniques; however, in order to describe the essence of the procedure, we here restrict ourselves to the case of the *local* pseudopotentials. The inclusion of nonlocal parts of the norm-conserving pseudopotentials is straightforward, which is given in the next chapter. Throughout this book we deal with only ballistic electronic transport. In general, the influence of phonon scattering (ionic vibrations) on electronic transport more weakens so as to be ignored at lower temperature, and then calculation results should be compared to low temperature measurements. For electron-phonon interaction effects, see Datta (1995).

For obtaining an exact theoretical knowledge on ballistic transport through a nanoscale junction, we have to treat the junction intervening between two *semi-infinite crystalline* electrodes, as indicated in Fig. 6.2. It is not until such a system is employed that physical quantities on electronic transport, such as electric conductance and current flow under steady states, can be most correctly described in terms of the transmission probability in the *scattering theory*. There are some problems to be overcome: Firstly, we need to deal with the system inherently losing the periodicity in the direction parallel to the junction. For such a system, the conventional repeated slab model employed in the plane-wave expansion approach breaks down due to the difficulty of incorporating nonperiodicity in it. The second problem is that global wave functions for *infinitely* extended states continuing from one side to the other are required to be accurately calculated.

In what follows, we consider the procedure for solving the Kohn–Sham equation in a system with nanostructures sandwiched between two semi-infinite crystalline electrodes. It is evident that an effective potential is close to periodic bulk potentials as it goes deeply inside the left and right electrodes, so that the whole infinite system can be appropriately divided into three parts: the left electrode, the transition region, and the right electrode (Fig. 6.2). The transition region should then be taken as being so large that the potentials on the left and right boundary planes in the transition region can have a sufficiently smooth continuation of the periodic potentials of the respective electrodes. In Fig. 6.2, $r_{||} = (x, y)$ and z coordinates are chosen to be perpendicular and parallel to the nanoscale junction, respectively. Here and hereafter we assume two-dimensional periodicity for the $r_{||}$ coordinates and nonperiodicity for the z one. By means of the present modeling, almost any infinitely extending system without periodicity, such as a tip-sample system (e.g., scanning tunneling microscopy) and an interface between two different bulks, as well as a nanoscale junction, can be properly treated.

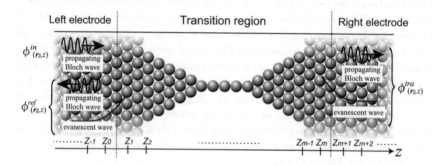

Fig. 6.2   Sketch of a system with the transition region sandwiched between the left and right semi-infinite crystalline electrodes. In the left electrode, the incident wave and the reflected waves consisting of the propagating and evanescent ones are illustrated by $\phi^{in}(r_{||}, z)$ and $\phi^{ref}(r_{||}, z)$, respectively, whereas in the right electrode, the transmitted waves composed of the propagating and decaying evanescent ones toward the right side are denoted by $\phi^{tra}(r_{||}, z)$. Here, $r_{||} = (x, y)$ and z are coordinates perpendicular and parallel to the nanoscale junction, respectively.

As known in the scattering theory, the solution of the Kohn–Sham equation for the whole system corresponds to a scattering wave function specified by a particular incident Bloch wave. The scattering wave function is classified into two types. One is incident from the left electrode with the reflection to the left electrode and transmission to the right electrode, which is depicted in Fig. 6.2. The other solution is vice versa, that is, incident from the right electrode with the reflection to the right electrode and transmission to the left electrode. Throughout this book we consider the former case, unless otherwise specified. Then, the asymptotic form of the scattering wave function consists of the incident wave $\phi^{in}(\boldsymbol{r}_{||}, z)$ plus reflected waves $\phi^{ref}(\boldsymbol{r}_{||}, z)$ in the left electrode and the transmitted waves $\phi^{tra}(\boldsymbol{r}_{||}, z)$ in the right electrode. We will construct below the global scattering wave function satisfying this asymptotic behavior by means of matching together the near-boundary values of the wave function, namely, its values near the right boundary of the left electrode and those near the left boundary of the right electrode, overbridging the transition region. Thus this matching procedure is referred to as the overbridging boundary-matching (OBM) method. For this purpose, we will introduce a Green's function carrying all of the information on the transition region.

The OBM method relies on the real-space finite-difference approach developed in Part I. In this approach, the real space is uniformly divided into the grid points $\{(x_i, y_j, z_k)\}$, $i = 1, 2, ..., N_x$, $j = 1, 2, ..., N_y$, and $k = -\infty, ..., -1, 0, 1, ..., \infty$, on which wave functions, electronic charge density, and potentials are directly defined. The numbers N_x and N_y are finite, since the system is assumed to be periodic in the x and y directions. In this chapter, we restrict our formulation to the case of adopting the central finite-difference formula for the kinetic-energy operator [$N_f = 1$, see Eq. (1.5)], for simplicity. The extension of the methodology to the case of a higher-order finite-difference formula ($N_f \geq 2$) is straightforward. A set of values of the wave function on the x–y plane at the $z = z_k$ point is treated as a columnar vector $\Psi(z_k)$ which has the N ($= N_f \times N_x \times N_y$) components of $\{\psi(x_i, y_j, z_k)\}$, where $i = 1, ..., N_x$ and $j = 1, ..., N_y$. The discretized Kohn–Sham equation is written as the three-term matrix equations,

$$-B(z_{k-1})^{\dagger}\Psi(z_{k-1}) + \left[E - H(z_k; \boldsymbol{k}_{||})\right]\Psi(z_k) - B(z_k)\Psi(z_{k+1}) = 0$$
$$(k = -\infty, ..., -1, 0, 1, ..., \infty), \quad (6.7)$$

where we have, in the case of the central finite difference [see Eq. (1.9)],

$$\cdots = B(z_{k-1}) = B(z_k) = B(z_{k+1}) = \cdots = -\frac{1}{2h_z^2}I \qquad (6.8)$$

with I being the N-dimensional unit matrix, and $H(z_k; \boldsymbol{k}_{||})$ denoting an N-dimensional diagonal matrix including the potential on the x–y plane at the $z = z_k$ point. Here, h_z is the grid spacing in the z direction and $\boldsymbol{k}_{||} = (k_x, k_y)$ is a lateral Bloch wave vector within the first Brillouin zone, i.e., $-\pi/L_\mu \leq k_\mu < \pi/L_\mu$ ($\mu = x, y$) with L_μ being the side lengths of the supercell of two-dimensional periodicity.

It is noted that the description of the Kohn–Sham equation in a relation of three adjacent terms is also possible in the Laue representation that employs a two-dimensional plane-wave expansion in the lateral directions and a real-space discretization for the z coordinate [for example, see Marcus and Jepsen (1968) and Hirose and Tsukada (1994, 1995)]. The OBM formulation discussed below is mostly applicable to it. However, from the practical point of view, the real-space finite-difference method has the following advantages compared with the method of the Laue representation:

(i) The finite differentiation for the kinetic-energy operator is treated on the equal footing in all the three directions. This avoids numerical errors due to the artificial anisotropy between the lateral and longitudinal directions for any value of the grid spacing.

(ii) Block matrices $H(z_k; \mathbf{k}_{||})$ and $B(z_k)$ are very sparse even when a higher-order finite-difference formula is adopted, which allows us to carry out the multiplications of these matrices and vectors $\Psi(z_k)$ with reduced computational efforts. Then, making use of iterative algorithms, for example, steepest-descent (SD) and conjugate-gradient (CG) algorithms, yields the savings of computer time and memories in solving the Kohn–Sham equation and calculating the Green's function matrix needed in the OBM method. Accordingly, the entirely real-space calculating method, not relying on fast Fourier transforms but making the most of calculation procedures developed in Part I, is more suitable for examining transport properties of large-sized systems using a massively parallel computer.

(iii) In the x and y directions, not only periodic boundary conditions but also isolated ones are available. In the latter case, electrodes may be in turn treated as leads. This way, the theory can be so extended as to include three- and multi-lead problems.

The OBM formulation can also be applicable to the tight-binding method, the overview of which is presented in Appendix B.

The discretized Kohn–Sham equation of the whole system is schematically shown in Fig. 6.3. The dotted lines represent the borders of the transition region. Each square denotes an N-dimensional block-matrix element of the Hamiltonian, namely, a diagonal element $[E - H(z_k; \mathbf{k}_{||})]$ and an off-diagonal element $-B(z_k)^{(\dagger)}$, and thus the Hamiltonian is described in the form of a block-tridiagonal matrix. In Fig. 6.3 the two squares shaded by oblique lines are the 'boundary matrices' $-B(z_{-1})^\dagger$ and $-B(z_{m+1})$ which correlate the boundary values $\Psi(z_0)$ and $\Psi(z_{m+1})$ with $\Psi(z_{-1})$ and $\Psi(z_{m+2})$, respectively.

We now attempt to drive a wave-function matching formula to construct the scattering wave function extending over the entire system. The Kohn–

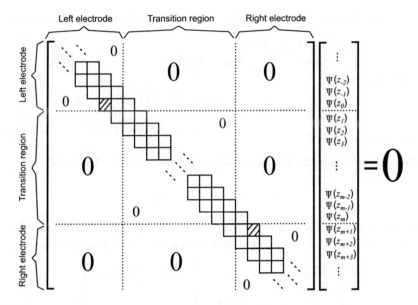

Fig. 6.3 Schematic representation of the discretized Kohn–Sham equation for the system illustrated in Fig. 6.2.

Sham equation is a second-order differential equation. The solution of a second-order differential equation can be determined by specifying a value and a first derivative at one point or by specifying values at two different points. We have adopted the latter. For these two specified values, we choose $\Psi(z_{-1})$ and $\Psi(z_{m+2})$ inside the left and right electrode regions, respectively (see Fig. 6.3). With the given values of $\Psi(z_{-1})$ and $\Psi(z_{m+2})$, the discretized Kohn–Sham equation for the area from $z = z_0$ to z_{m+1} is expressed as the simultaneous linear equations with respect to $\{\Psi(z_k)\}$, $k = 0, 1, ..., m + 1$,

$$\left[E - \hat{H}_T(\boldsymbol{k}_{||})\right] \begin{bmatrix} \Psi(z_0) \\ \Psi(z_1) \\ \vdots \\ \Psi(z_m) \\ \Psi(z_{m+1}) \end{bmatrix} = \begin{bmatrix} B(z_{-1})^\dagger \Psi(z_{-1}) \\ 0 \\ \vdots \\ 0 \\ B(z_{m+1})\Psi(z_{m+2}) \end{bmatrix}, \qquad (6.9)$$

where E is a Kohn–Sham energy, and $\hat{H}_T(\boldsymbol{k}_{||})$ is the Hamiltonian of a truncated part of the system sandwiched between the planes at $z = z_0$ and

z_{m+1}, which forms the $(m + 2)N$-dimensional block-tridiagonal matrix of

$$\hat{H}_T(\boldsymbol{k}_{||}) = \begin{bmatrix} H(z_0; \boldsymbol{k}_{||}) & B(z_0) & & & 0 \\ B(z_0)^\dagger & H(z_1; \boldsymbol{k}_{||}) & B(z_1) & & \\ & \ddots & \ddots & \ddots & \\ & & B(z_{m-1})^\dagger & H(z_m; \boldsymbol{k}_{||}) & B(z_m) \\ 0 & & & B(z_m)^\dagger & H(z_{m+1}; \boldsymbol{k}_{||}) \end{bmatrix}. \tag{6.10}$$

We now introduce the Green's function matrix associated with $\hat{H}_T(\boldsymbol{k}_{||})$:

$$\hat{\mathcal{G}}_T(E, \boldsymbol{k}_{||}) = \left[E - \hat{H}_T(\boldsymbol{k}_{||}) \right]^{-1}. \tag{6.11}$$

When the matrix $\hat{\mathcal{G}}_T(E, \boldsymbol{k}_{||})$ is known for a given energy E and lateral Bloch wave vector $\boldsymbol{k}_{||}$, one can immediately obtain the solution of (6.9) as

$$\begin{bmatrix} \Psi(z_0) \\ \Psi(z_1) \\ \vdots \\ \Psi(z_m) \\ \Psi(z_{m+1}) \end{bmatrix} = \hat{\mathcal{G}}_T(E, \boldsymbol{k}_{||}) \begin{bmatrix} B(z_{-1})^\dagger \Psi(z_{-1}) \\ 0 \\ \vdots \\ 0 \\ B(z_{m+1}) \Psi(z_{m+2}) \end{bmatrix}. \tag{6.12}$$

Thus, it can be seen that all the values of the wave function in the transition region, $\Psi(z_1)$, $\Psi(z_2)$,..., and $\Psi(z_m)$, are directly connected to the two values of $\Psi(z_{-1})$ and $\Psi(z_{m+2})$ by means of the matrix $\hat{\mathcal{G}}_T(E, \boldsymbol{k}_{||})$, and that the values of the wave function outside the transition region, $\Psi(z_{-1})$, $\Psi(z_0)$, $\Psi(z_{m+1})$ and $\Psi(z_{m+2})$, have the simple relation of

$$\begin{bmatrix} \Psi(z_0) \\ \Psi(z_{m+1}) \end{bmatrix} = \begin{bmatrix} \mathcal{G}_T(z_0, z_0) & \mathcal{G}_T(z_0, z_{m+1}) \\ \mathcal{G}_T(z_{m+1}, z_0) & \mathcal{G}_T(z_{m+1}, z_{m+1}) \end{bmatrix} \begin{bmatrix} B(z_{-1})^\dagger \Psi(z_{-1}) \\ B(z_{m+1}) \Psi(z_{m+2}) \end{bmatrix}, \tag{6.13}$$

where $\mathcal{G}_T(z_k, z_l)$ is the N-dimensional (k, l) (i.e., the k-th row and l-th column) block-matrix element of the Green's function matrix $\hat{\mathcal{G}}_T(E, \boldsymbol{k}_{||})$. The dependence of $\mathcal{G}_T(z_k, z_l)$ on E and $\boldsymbol{k}_{||}$ is ignored for simplicity. Equation (6.13) is a realization of the OBM formula that connects the values near the boundary of the left electrode, $\Psi(z_0)$ and $\Psi(z_{-1})$, and those of the right electrode, $\Psi(z_{m+1})$ and $\Psi(z_{m+2})$, in which the near-boundary values of the wave function in the left and right electrodes are directly connected with each other by the Green's function matrix overlying the transition region.

Since the effective potentials in the left and right semi-infinite crystalline electrodes tend to take a periodic property deep inside the electrodes due to the screening effect, it is appropriate to describe the global scattering wave function over the entire system in terms of a linear combination of the solutions of the Kohn–Sham equation under the periodic potentials inside the left and right electrodes. We refer to such solutions inside semi-infinite crystalline bulks as generalized Bloch states (see the next section for detailed discussions of generalized Bloch states). As illustrated in Fig. 6.2, $\phi^{in}(r_{||}, z)$ is an incident Bloch wave propagating from deep inside the left electrode, and $\phi^{ref}(r_{||}, z)$ is a set of reflected waves that propagate and decay into the left electrode; $\phi^{tra}(r_{||}, z)$ within the right electrode is a set of transmitted waves, i.e., an ensemble of rightward propagating Bloch waves and decaying ones toward the right side. Decaying (or growing) waves, which behave like exponential functions, are called evanescent waves, and both propagating Bloch waves and evanescent ones constitute generalized Bloch states inside the semi-infinite electrodes (see the next section). In the left-electrode region, the scattering wave function $\Psi(z_k)$ is expressed as a linear combination of $\phi^{in}(r_{||}, z)$ and $\phi^{ref}(r_{||}, z)$,

$$\Psi(z_k) = \Phi^{in}(z_k) + \sum_{i=1}^{N} r_i \Phi_i^{ref}(z_k) \quad (k \leq 0), \tag{6.14}$$

while in the right electrode region, it is given by a linear combination of a set of transmitted waves $\phi^{tra}(r_{||}, z)$,

$$\Psi(z_k) = \sum_{i=1}^{N} t_i \Phi_i^{tra}(z_k) \quad (k \geq m+1). \tag{6.15}$$

Here, r_i and t_i $(i = 1, ..., N)$ are unknown reflection and transmission coefficients, respectively, and $\Phi^A(z_k)$ $(A = in, ref$ and $tra)$ are N-dimensional columnar vectors constructed by $\{\phi^A(x_i, y_j, z_k)\}$, where $i = 1, 2, ..., N_x$ and $j = 1, 2, ..., N_y$, on the x–y plane at the $z = z_k$ point. We show in Section 6.5 that the numbers of both $\Phi_i^{ref}(z_k)$ in (6.14) and $\Phi_i^{tra}(z_k)$ in (6.15) are equally N. Introducing the coefficient vectors

$$X^{ref} = [r_1, r_2, ..., r_N]^t \quad \text{and} \quad X^{tra} = [t_1, t_2, ..., t_N]^t \tag{6.16}$$

and the N-dimensional matrices constructed of generalized Bloch states

$$Q^A(z_k) = \left[\Phi_1^A(z_k), \Phi_2^A(z_k), ..., \Phi_N^A(z_k) \right] \quad (A = ref \text{ and } tra), \tag{6.17}$$

we rewrite (6.14) and (6.15) as

$$\Psi(z_k) = \Phi^{in}(z_k) + Q^{ref}(z_k)X^{ref} \quad (k \leq 0) \tag{6.18}$$

and

$$\Psi(z_k) = Q^{tra}(z_k)X^{tra} \quad (k \geq m+1). \tag{6.19}$$

Thus, substituting (6.18) for $k = -1, 0$ and (6.19) for $k = m+1, m+2$ into (6.13), we have the simultaneous linear equations with regard to the coefficient vectors X^{ref} and X^{tra},

$$\begin{bmatrix} \mathcal{W}_{0,0}Q^{ref}(z_{-1}) - Q^{ref}(z_0) & \mathcal{W}_{0,m+1}Q^{tra}(z_{m+2}) \\ \mathcal{W}_{m+1,0}Q^{ref}(z_{-1}) & \mathcal{W}_{m+1,m+1}Q^{tra}(z_{m+2}) - Q^{tra}(z_{m+1}) \end{bmatrix} \begin{bmatrix} X^{ref} \\ X^{tra} \end{bmatrix}$$
$$= -\begin{bmatrix} \mathcal{W}_{0,0}\Phi^{in}(z_{-1}) - \Phi^{in}(z_0) \\ \mathcal{W}_{m+1,0}\Phi^{in}(z_{-1}) \end{bmatrix}, \tag{6.20}$$

where

$$\begin{aligned} \mathcal{W}_{0,0} &= \mathcal{G}_T(z_0, z_0)B(z_{-1})^\dagger \\ \mathcal{W}_{0,m+1} &= \mathcal{G}_T(z_0, z_{m+1})B(z_{m+1}) \\ \mathcal{W}_{m+1,0} &= \mathcal{G}_T(z_{m+1}, z_0)B(z_{-1})^\dagger \\ \mathcal{W}_{m+1,m+1} &= \mathcal{G}_T(z_{m+1}, z_{m+1})B(z_{m+1}). \end{aligned} \tag{6.21}$$

From (6.20) we can determine all of the coefficients r_i and t_i in principle, if the generalized Bloch states Φ, and therefore Q, are known. However, Q in this matching formula always includes exponentially growing and decaying evanescent waves, and in some cases, the inclusion of these evanescent waves gives rise to a numerical problem such that the smaller the grid spacing h_μ ($\mu = x, y$), the more serious the numerical errors (in the next section, we derive a formula for calculating the generalized Bloch states Φ and Q, and discuss the numerical accuracy of the formula in detail).

In order to overcome this numerical difficulty, we introduce the following ratios of the generalized Bloch states:

$$\begin{aligned} R^{ref}(z_0) &= Q^{ref}(z_{-1})Q^{ref}(z_0)^{-1} \\ R^{tra}(z_{m+2}) &= Q^{tra}(z_{m+2})Q^{tra}(z_{m+1})^{-1}. \end{aligned} \tag{6.22}$$

The ratio matrices $R^{ref}(z_0)$ and $R^{tra}(z_{m+2})$ do not involve the values of evanescent wave functions themselves, but include the ratios of the values of evanescent wave functions on the planes at two neighboring z points, which correspond to logarithmic derivatives in the continuous limit of real space [Kobayashi (1999b)]; thus, numerical errors due to the appearance of evanescent waves do not accumulate during the computing of the ratio matrices $R^{ref}(z_0)$ and $R^{ref}(z_{m+2})$. The method of calculating the ratio matrices directly, not via the values of evanescent wave functions themselves, is given in Section 6.4.

We rewrite (6.20) in terms of $R^{ref}(z_0)$ and $R^{tra}(z_{m+2})$ as

$$\begin{bmatrix} \mathcal{W}_{0,0} - R^{ref}(z_0)^{-1} & \mathcal{W}_{0,m+1} \\ \mathcal{W}_{m+1,0} & \mathcal{W}_{m+1,m+1} - R^{tra}(z_{m+2})^{-1} \end{bmatrix} \begin{bmatrix} Q^{ref}(z_{-1})X^{ref} \\ Q^{tra}(z_{m+2})X^{tra} \end{bmatrix}$$
$$= - \begin{bmatrix} \mathcal{W}_{0,0}\Phi^{in}(z_{-1}) - \Phi^{in}(z_0) \\ \mathcal{W}_{m+1,0}\Phi^{in}(z_{-1}) \end{bmatrix}, \qquad (6.23)$$

and using the equations derived from (6.18) and (6.19), namely,

$$Q^{ref}(z_{-1})X^{ref} = \Psi(z_{-1}) - \Phi^{in}(z_{-1})$$
$$Q^{tra}(z_{m+2})X^{tra} = \Psi(z_{m+2}), \qquad (6.24)$$

we obtain the OBM formula which matches the values of the wave function, $\Psi(z_{-1})$ and $\Psi(z_{m+2})$:

$$\begin{bmatrix} \mathcal{W}_{0,0} - R^{ref}(z_0)^{-1} & \mathcal{W}_{0,m+1} \\ \mathcal{W}_{m+1,0} & \mathcal{W}_{m+1,m+1} - R^{tra}(z_{m+2})^{-1} \end{bmatrix} \begin{bmatrix} \Psi(z_{-1}) \\ \Psi(z_{m+2}) \end{bmatrix}$$
$$= - \begin{bmatrix} R^{ref}(z_0)^{-1}\Phi^{in}(z_{-1}) - \Phi^{in}(z_0) \\ 0 \end{bmatrix}. \qquad (6.25)$$

It should be emphasized that an incident wave Φ^{in} in the right-hand side of (6.25) represents only the *propagating* Bloch wave so that no evanescent waves themselves take their places in (6.25), and eventually, the numerical problem caused by exponentially growing or decaying evanescent waves is completely excluded in (6.25). We note that $\Psi(z_{-1})$ and $\Psi(z_{m+2})$ are solved as the solutions of (6.25) for each incident propagating wave Φ^{in}, and once they are obtained, all the values of the scattering wave function around the transition region, $\Psi(z_k)$ $(k = 0, ..., m+1)$, are determined from (6.12), in which only a part of the block matrix elements of the Green's function matrix, $\mathcal{G}_T(z_k, z_l)$ $(k = 0, 1, ..., m+1,$ and $l = 0, m+1)$, namely, the left and right block-columns of $\hat{\mathcal{G}}_T(E, \mathbf{k}_{||})$ $(= [E - \hat{H}_T(\mathbf{k}_{||})]^{-1})$, are needed. In calculating these block columns, the direct inversion of $[E - \hat{H}_T(\mathbf{k}_{||})]$ may be a computationally hard task when the transition region is of a large-sized structure; instead, one can make efficient use of usual SD and CG iterative algorithms, since $\hat{H}_T(\mathbf{k}_{||})$ is Hermitian and these algorithms are known to quickly converge for an Hermitian matrix.

Finally, we give the OBM formula with respect to $\Psi(z_0)$ and $\Psi(z_{m+1})$,

$$\begin{bmatrix} \mathcal{W}_{0,0}R^{ref}(z_0) - I & \mathcal{W}_{0,m+1}R^{tra}(z_{m+2}) \\ \mathcal{W}_{m+1,0}R^{ref}(z_0) & \mathcal{W}_{m+1,m+1}R^{tra}(z_{m+2}) - I \end{bmatrix} \begin{bmatrix} \Psi(z_0) \\ \Psi(z_{m+1}) \end{bmatrix}$$
$$= - \begin{bmatrix} \mathcal{W}_{0,0} \\ \mathcal{W}_{m+1,0} \end{bmatrix} (\Phi^{in}(z_{-1}) - R^{ref}(z_0)\Phi^{in}(z_0)). \qquad (6.26)$$

This formula is derived from (6.25) using the relations,

$$\Psi(z_{-1}) - \Phi^{in}(z_{-1}) = R^{ref}(z_0) \left[\Psi(z_0) - \Phi^{in}(z_0)\right]$$
$$\Psi(z_{m+2}) = R^{tra}(z_{m+2})\Psi(z_{m+1}) \tag{6.27}$$

which come from (6.18), (6.19) and (6.22). We discuss (6.26) with relation to the Green's function matching formula (9.91) in Chapter 9, where the matching problem is reformulated in terms of the Green's function matrix associated with an infinitely extending whole system.

Other methodologies for determining scattering wave functions of infinite (or semi-infinite) systems are seen in the literature; for example, Bennett and Duke (1967), Marcus and Jepsen (1968), Lang and Kohn (1970), Appelbaum and Hamann (1972), Lang and Williams (1975), Vigneron and Lambin (1979), Lee and Joannopoulos (1981a), Ferrante and Smith (1985), Stiles and Hamann (1988), Sautet and Joachim (1988), Lent and Kirkner (1990), Takagaki and Ferry (1992), Doyen *et al.* (1993), Zhu *et al.* (1994), Hirose and Tsukada (1994, 1995), Sheng and Xia (1996), Mozos *et al.* (1997), Hummel and Bross (1998), Emberly and Kirczenow (1998, 1999), Kobayashi (1999b), Choi and Ihm (1999), Gohda *et al.* (2000), Wortmann *et al.* (2002) and Ami *et al.* (2003).

6.3　Computational Scheme for Generalized Bloch States
I. The Generalized Eigenvalue Equation

In exhibiting the wave-function matching formulas (6.20), (6.25) and (6.26), we have regarded the generalized Bloch states Φ^A ($A = in, ref$ and tra), namely, the solutions of the Kohn–Sham equation deep inside semi-infinite crystalline electrodes, as given. We present here the procedure for calculating them on the basis of the OBM scheme.

The potentials deep inside electrodes are assumed to be periodic along the z direction as well as the x and y directions. Let us consider an infinite crystalline bulk (Fig. 6.4). Figure 6.5 schematically shows the discretized Kohn–Sham equation (6.7) for such a crystalline bulk, the Hamiltonian of which is described by an infinite block-tridiagonal matrix, where squares represent N-dimensional block-matrix elements, and the two squares shaded by oblique lines are the 'boundary matrices', $-B(z_m^M)^\dagger$ and $-B(z_m^M)$ [note that $B(z_m^{M-1}) = B(z_m^M)$ due to the periodicity]. For the given values of $\Phi(z_m^{M-1})$ and $\Phi(z_1^{M+1})$, the Kohn–Sham equation in the M-th unit cell forms the simultaneous linear equations with respect to $\{\Phi(z_k^M)\}$, $k = 1, 2, ..., m$:

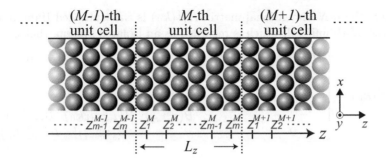

Fig. 6.4   Schematic view of a crystalline bulk. z_k^M represents the z coordinate at the k-th grid point in the M-th unit cell. L_z denotes the unit-cell length in the z direction.

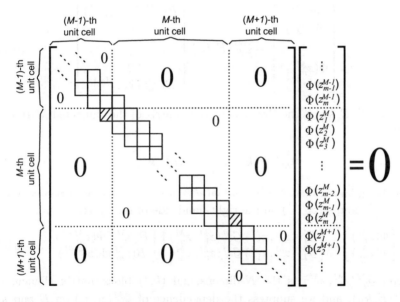

Fig. 6.5   Schematic representation of the discretized Kohn–Sham equation for the crystalline bulk illustrated in Fig. 6.4. The dotted lines fix the boundaries of the M-th unit cell with two adjacent unit cells.

$$\left[E - \hat{H}_T^M(\mathbf{k}_\parallel)\right]\begin{bmatrix} \Phi(z_1^M) \\ \Phi(z_2^M) \\ \vdots \\ \Phi(z_{m-1}^M) \\ \Phi(z_m^M) \end{bmatrix} = \begin{bmatrix} B(z_m^M)^\dagger \Phi(z_m^{M-1}) \\ 0 \\ \vdots \\ 0 \\ B(z_m^M)\Phi(z_1^{M+1}) \end{bmatrix}, \tag{6.28}$$

where the mN-dimensional matrix $\hat{H}_T^M(\boldsymbol{k}_{||})$ is the truncated Hamiltonian of the M-th unit cell which is regarded as an isolated system, that is,

$$\hat{H}_T^M(\boldsymbol{k}_{||}) = \begin{bmatrix} H(z_1^M;\boldsymbol{k}_{||}) & B(z_1^M) & & & 0 \\ B(z_1^M)^\dagger & H(z_2^M;\boldsymbol{k}_{||}) & B(z_2^M) & & \\ & \ddots & \ddots & \ddots & \\ & & B(z_{m-2}^M)^\dagger & H(z_{m-1}^M;\boldsymbol{k}_{||}) & B(z_{m-1}^M) \\ 0 & & & B(z_{m-1}^M)^\dagger & H(z_m^M;\boldsymbol{k}_{||}) \end{bmatrix}. \tag{6.29}$$

Here, $H(z_k^M;\boldsymbol{k}_{||})$ represents a block-diagonal element including the bulklike potential on the x–y plane at the z_k^M point. The wave function in the M-th unit cell is then obtained by solving (6.28) as

$$\begin{bmatrix} \Phi(z_1^M) \\ \Phi(z_2^M) \\ \vdots \\ \Phi(z_{m-1}^M) \\ \Phi(z_m^M) \end{bmatrix} = \hat{\mathcal{G}}_T^M(E,\boldsymbol{k}_{||}) \begin{bmatrix} B(z_m^M)^\dagger \Phi(z_m^{M-1}) \\ 0 \\ \vdots \\ 0 \\ B(z_m^M)\Phi(z_1^{M+1}) \end{bmatrix}, \tag{6.30}$$

where $\hat{\mathcal{G}}_T^M(E,\boldsymbol{k}_{||})$ is the Green's function matrix of the truncated M-th unit cell,

$$\hat{\mathcal{G}}_T^M(E,\boldsymbol{k}_{||}) = \left[E - \hat{H}_T^M(\boldsymbol{k}_{||})\right]^{-1}. \tag{6.31}$$

The boundary values, $\Phi(z_1^M)$ and $\Phi(z_m^M)$, are connected with two values of $\Phi(z_m^{M-1})$ and $\Phi(z_1^{M+1})$ on the right-hand side of (6.29) as

$$\begin{bmatrix} \Phi(z_1^M) \\ \Phi(z_m^M) \end{bmatrix} = \begin{bmatrix} \mathcal{G}_T^M(z_1^M,z_1^M) & \mathcal{G}_T^M(z_1^M,z_m^M) \\ \mathcal{G}_T^M(z_m^M,z_1^M) & \mathcal{G}_T^M(z_m^M,z_m^M) \end{bmatrix} \begin{bmatrix} B(z_m^M)^\dagger \Phi(z_m^{M-1}) \\ B(z_m^M)\Phi(z_1^{M+1}) \end{bmatrix}. \tag{6.32}$$

Here, $\mathcal{G}_T^M(z_k^M,z_l^M)$ is the N-dimensional (k,l) block-matrix element of $\hat{\mathcal{G}}_T^M(E,\boldsymbol{k}_{||})$, and we suppress the dependence of $\mathcal{G}_T^M(z_k,z_l)$ on E and $\boldsymbol{k}_{||}$ for simplicity.

The other restriction that the wave function Φ has to satisfy is the following Bloch condition attributed to the periodicity of the potential with the length L_z along the z direction:

$$\Phi(z + L_z) = \lambda \Phi(z), \tag{6.33}$$

where $\lambda = e^{ik_z L_z}$, and k_z is the z component of a Bloch wave vector. In contrast to the case of usual band calculations, we here choose k_z to be a *complex* number in general; thereby we can deal with a *generalized* Bloch state Φ. Thus we refer to (6.33) as the generalized Bloch condition. Now according to (6.33), $\Phi(z_m^M)$ and $\Phi(z_1^{M+1})$ are related to $\Phi(z_m^{M-1})$ and $\Phi(z_1^M)$ as

$$\Phi(z_m^M) = \lambda\Phi(z_m^{M-1}) \quad \text{and} \quad \Phi(z_1^{M+1}) = \lambda\Phi(z_1^M). \tag{6.34}$$

Consequently, eliminating $\Phi(z_1^M)$ and $\Phi(z_m^M)$ from (6.32) by using (6.34), we obtain the following Theorem:

Theorem 6.1 *The generalized Bloch state with a complex Bloch wave-vector component k_z obeys the 2N-dimensional generalized eigenvalue equation with an eigenvalue $\lambda = e^{ik_z L_z}$:*

$$\Pi_1(E, \boldsymbol{k}_{||}) \begin{bmatrix} \Phi(z_m^{M-1}) \\ \Phi(z_1^{M+1}) \end{bmatrix} = \lambda\Pi_2(E, \boldsymbol{k}_{||}) \begin{bmatrix} \Phi(z_m^{M-1}) \\ \Phi(z_1^{M+1}) \end{bmatrix}, \tag{6.35}$$

where

$$\Pi_1(E, \boldsymbol{k}_{||}) = \begin{bmatrix} \mathcal{W}_{m,1}^M & \mathcal{W}_{m,m}^M \\ 0 & I \end{bmatrix}$$

$$\Pi_2(E, \boldsymbol{k}_{||}) = \begin{bmatrix} I & 0 \\ \mathcal{W}_{1,1}^M & \mathcal{W}_{1,m}^M \end{bmatrix}, \tag{6.36}$$

and

$$\mathcal{W}_{1,1}^M = \mathcal{G}_T^M(z_1^M, z_1^M) B(z_m^M)^\dagger$$
$$\mathcal{W}_{1,m}^M = \mathcal{G}_T^M(z_1^M, z_m^M) B(z_m^M)$$
$$\mathcal{W}_{m,1}^M = \mathcal{G}_T^M(z_m^M, z_1^M) B(z_m^M)^\dagger$$
$$\mathcal{W}_{m,m}^M = \mathcal{G}_T^M(z_m^M, z_m^M) B(z_m^M). \tag{6.37}$$

Analogous expressions of the generalized eigenvalue equation are found in Wachutka (1986), Stiles and Hamann (1988), Hummel and Bross (1998), Kobayashi (1999b) and Choi and Ihm (1999).

In principle, the generalized Bloch state Φ and Bloch factor λ ($= e^{ik_z L_z}$) are determined from (6.35) as its eigenvector and eigenvalue, respectively, even though in practice a numerical problem frequently arises, as will be discussed later. The eigenvectors of (6.35) are classified into two different types according to whether $|\lambda| = 1$ or $|\lambda| \neq 1$. The eigenvectors with $|\lambda| = 1$ correspond to ordinary propagating Bloch waves, and k_z is confined to a real number. When $|\lambda| \neq 1$, the eigenvectors correspond to evanescent waves, and k_z is extended to the field of a complex number; more precisely, the eigenvectors with $|\lambda| > 1$ ($|\lambda| < 1$) represent leftward (rightward) decreasing evanescent waves, as seen by the definition of λ, (6.33). Propagating and evanescent waves make up the *complex* band structure. Generally, the term 'complex band structure' refers to energy bands consisting of the so-lutions $E(\boldsymbol{k})$ with real energy E and complex Bloch wave vectors $\boldsymbol{k}$, in analogy to the conventional (real $\boldsymbol{k}$) band structure. For the complex band structure, see, for example, Wachutka (1986), Stiles and Hamann (1988) and Wortmann *et al.* (2002).

Propagating Bloch waves are physical solutions of the Kohn–Sham equation (6.7) for an infinite system with a periodic bulklike potential, while evanescent waves are not physical solutions but simply mathematical ones for such an infinite system, because they behave like exponentially increasing and decreasing functions and are thus unnormalizable in the whole space; however, leftward (rightward) decreasing evanescent waves are obviously physical inside the left (right) semi-infinite electrode, so that we have to take them into account as ingredients for describing the global scattering solution, when treating the system with the transition region intervening between these electrodes as illustrated in Fig. 6.2.

It is noted that in solving (6.35), a Kohn–Sham energy E as well as a lateral wave vector $\boldsymbol{k}_{||}$ is dealt with as an input parameter, and a Bloch wave-vector component k_z as an output one, which is contrary to usual band-structure calculations for crystalline bulks using the supercell technique; in these usual calculations, one works out the standard eigenvalue equation

$$\hat{H}_{per}(\omega = e^{ik_z L_z}, \boldsymbol{k}_{||})\hat{\Phi} = E\hat{\Phi} \tag{6.38}$$

of the mN-dimensional *periodic* Hamiltonian of the M-th unit cell, $\hat{H}_{per}^M(\omega = e^{ik_z L_z}, \boldsymbol{k}_{||})$, which is expressed in the form of

$$\hat{H}_{per}^M(\omega, \boldsymbol{k}_{||})$$

$$= \begin{bmatrix} H(z_1^M; \boldsymbol{k}_{||}) & B(z_1^M) & 0 & \cdots & 0 & \omega^{-1}B(z_m^M)^{\dagger} \\ B(z_1^M)^{\dagger} & H(z_2^M; \boldsymbol{k}_{||}) & B(z_2^M) & & & 0 \\ 0 & \ddots & \ddots & \ddots & & \vdots \\ \vdots & & \ddots & \ddots & \ddots & 0 \\ 0 & & & B(z_{m-2}^M)^{\dagger} & H(z_{m-1}^M; \boldsymbol{k}_{||}) & B(z_{m-1}^M) \\ \omega B(z_m^M) & 0 & \cdots & 0 & B(z_{m-1}^M)^{\dagger} & H(z_m^M; \boldsymbol{k}_{||}) \end{bmatrix},$$

$$\tag{6.39}$$

taking a real k_z in the first Brillouin zone as an input parameter and E as an output one. It is worthwhile observing the 'equivalence' between the two eigenvalue problems, (6.35) and (6.38); the following theorem indicates that if

$$|E - \hat{H}_{per}^M(\omega, \boldsymbol{k}_{||})| = 0, \tag{6.40}$$

then

$$|\Pi_1(E, \mathbf{k}_{||}) - \omega\Pi_2(E, \mathbf{k}_{||})| = 0, \tag{6.41}$$

and vice versa, provided that E coincides with no eigenvalues of $\hat{H}_T^M(\mathbf{k}_{||})$ defined by (6.29). Here, the determinant of matrix A is denoted by $|A|$.

Theorem 6.2

$$\left|E - \hat{H}_T^M(\mathbf{k}_{||})\right|\left|\Pi_1(E, \mathbf{k}_{||}) - \omega\Pi_2(E, \mathbf{k}_{||})\right| = (-\omega)^N \left|E - \hat{H}_{per}^M(\omega, \mathbf{k}_{||})\right| \tag{6.42}$$

This formula holds for an arbitrary complex number ω. [Note that $\hat{H}_{per}^M$ of (6.39) is a 'physical' Hermitian Hamiltonian only when ω is a phase factor satisfying $|\omega| = 1$.]

Proof. From (6.35) – (6.37), one has

$$|\Pi_1(E, \mathbf{k}_{||}) - \omega\Pi_2(E, \mathbf{k}_{||})| = \begin{vmatrix} \mathcal{W}_{m,1}^M - \omega I & \mathcal{W}_{m,m}^M \\ -\omega\mathcal{W}_{1,1}^M & I - \omega\mathcal{W}_{1,m}^M \end{vmatrix}. \tag{6.43}$$

On the other hand, $|E - \hat{H}_{per}^M(\omega, \mathbf{k}_{||})|$ is written as

$$\begin{aligned}
|E - \hat{H}_{per}^M(\omega, \mathbf{k}_{||})| &= |E - \hat{H}_T^M(\mathbf{k}_{||}) - \delta\hat{H}| \\
&= |E - \hat{H}_T^M(\mathbf{k}_{||})||I - \hat{\mathcal{G}}_T^M(E, \mathbf{k}_{||})\delta\hat{H}|.
\end{aligned} \tag{6.44}$$

Here,

$$\begin{aligned}
\delta\hat{H} &= \hat{H}_{per}^M(\omega, \mathbf{k}_{||}) - \hat{H}_T^M(\mathbf{k}_{||}) \\
&= \begin{bmatrix} 0 & 0 \cdots 0 & \omega^{-1}B(z_m^M)^\dagger \\ 0 & 0 \quad 0 & 0 \\ \vdots & \vdots \quad \vdots & \vdots \\ 0 & 0 \quad 0 & 0 \\ \omega B(z_m^M) & 0 \cdots 0 & 0 \end{bmatrix},
\end{aligned} \tag{6.45}$$

and $\hat{\mathcal{G}}_T^M(E, \mathbf{k}_{||})$ is given by (6.31), i.e.,

$$\hat{\mathcal{G}}_T^M(E, \mathbf{k}_{||}) = \left[E - \hat{H}_T^M(\mathbf{k}_{||})\right]^{-1}. \tag{6.46}$$

Then one can easily find

$$|I - \hat{\mathcal{G}}_T^M(E, \boldsymbol{k}_{||})\delta\hat{H}| = \begin{vmatrix} I - \omega\mathcal{W}_{1,m}^M & 0 & \cdots & 0 & -\omega^{-1}\mathcal{W}_{1,1}^M \\ -\omega\mathcal{W}_{2,m}^M & I & \cdots & 0 & -\omega^{-1}\mathcal{W}_{2,1}^M \\ -\omega\mathcal{W}_{3,m}^M & 0 & I & \vdots & -\omega^{-1}\mathcal{W}_{3,1}^M \\ \vdots & & & \ddots & \vdots \\ & & \vdots & & I \\ -\omega\mathcal{W}_{m,m}^M & 0 & \cdots & 0 & I - \omega^{-1}\mathcal{W}_{m,1}^M \end{vmatrix}$$

$$= \begin{vmatrix} I - \omega\mathcal{W}_{1,m}^M & -\omega^{-1}\mathcal{W}_{1,1}^M \\ -\omega\mathcal{W}_{m,m}^M & I - \omega^{-1}\mathcal{W}_{m,1}^M \end{vmatrix}$$

$$= (-\omega)^{-N}\begin{vmatrix} I - \omega\mathcal{W}_{1,m}^M & \mathcal{W}_{1,1}^M \\ -\omega\mathcal{W}_{m,m}^M & \mathcal{W}_{m,1}^M - \omega I \end{vmatrix}, \qquad (6.47)$$

where for $k = 1, 2, ..., m$,

$$\mathcal{W}_{k,m}^M = \mathcal{G}_T^M(z_k^M, z_m^M)B(z_m^M) \quad \text{and} \quad \mathcal{W}_{k,1}^M = \mathcal{G}_T^M(z_k^M, z_1^M)B(z_m^M)^\dagger \quad (6.48)$$

with $\mathcal{G}_T^M(z_k^M, z_l^M)$ being the N-dimensional (k, l) block-matrix element of $\hat{\mathcal{G}}_T(E, \boldsymbol{k}_{||})$. Finally, by using the identity

$$\begin{vmatrix} A & B \\ \alpha C & D \end{vmatrix} = \begin{vmatrix} D & C \\ \alpha B & A \end{vmatrix}, \qquad (6.49)$$

where A, B, C, D are $N \times N$ submatrices and α is a number, one obtains (6.42) from (6.43), (6.44) and (6.47). □

A similar formula to (6.42) in Theorem 6.2 has been proposed by Lee and Joannopoulos (1981b) for a very restricted case. Equation (6.42) is a generalization of their formula and is applicable to any wider class of models; indeed, (6.42) can be verified without assuming a block-tridiagonal form for $\hat{H}_T^M$ and $\hat{H}_{per}^M$. Mainly relying on (6.42) with a real energy E replaced by a complex number Z $(= E + i\eta)$, we derive later Theorems 6.5 and 9.1, which provide the foundation for the OBM formalism.

For later use, we give some useful formulas (i) – (iii) associated with the generalized eigenvalue equation (6.35). To this end, we introduce a $2N$-dimensional matrix K by

$$K = \begin{bmatrix} 0 & -B(z_m^M) \\ B(z_m^M)^\dagger & 0 \end{bmatrix}, \qquad (6.50)$$

which is an anti-Hermitian matrix satisfying $K^\dagger = -K$, and utilize the identity easily derived from the definitions of Π_1 and Π_2 of (6.36),

$$\Pi_1^\dagger K \Pi_1 = \Pi_2^\dagger K \Pi_2$$
$$= \begin{bmatrix} 0 & -B(z_m^M)\mathcal{G}_T^M(z_1^M, z_m^M)B(z_m^M) \\ (B(z_m^M)\mathcal{G}_T^M(z_1^M, z_m^M)B(z_m^M))^\dagger & 0 \end{bmatrix},$$
$$(6.51)$$

where Π_i is the abbreviation of $\Pi_i(E, \boldsymbol{k}_{\parallel})$ $(i = 1, 2)$. Hereafter we denote (6.51) by

$$\tilde{K} = \Pi_1^\dagger K \Pi_1 = \Pi_2^\dagger K \Pi_2, \tag{6.52}$$

which is also anti-Hermitian, $\tilde{K}^\dagger = -\tilde{K}$.

Theorem 6.3

(i) *If λ $(\neq 0)$ is an eigenvalue of (6.35), then λ^{*-1} is also an eigenvalue of (6.35).*

(ii) *If $\mathcal{U}_a$ and $\mathcal{U}_b$ are solutions of (6.35) such that*

$$\Pi_1\mathcal{U}_a = \lambda_a\Pi_2\mathcal{U}_a \quad \text{and} \quad \Pi_1\mathcal{U}_b = \lambda_b\Pi_2\mathcal{U}_b, \tag{6.53}$$

and in addition $\lambda_a \neq \lambda_b^{*-1}$, then $\mathcal{U}_a$ and $\mathcal{U}_b$ satisfy the following conjugate relation with each other:

$$\mathcal{U}_a^\dagger \tilde{K} \mathcal{U}_b = \mathcal{U}_b^\dagger \tilde{K} \mathcal{U}_a = 0. \tag{6.54}$$

(iii) *If λ $(\neq 0)$ and $\mathcal{U}$ satisfy (6.35), then $\tilde{\mathcal{U}}$ defined by*

$$\tilde{\mathcal{U}} = K \Pi_2 \mathcal{U} \tag{6.55}$$

satisfies the eigenvalue equation of the corresponding adjoint matrices,

$$\Pi_1^\dagger \tilde{\mathcal{U}} = \lambda^{-1}\Pi_2^\dagger \tilde{\mathcal{U}}, \tag{6.56}$$

or equivalently,

$$\tilde{\mathcal{U}}^\dagger \Pi_1 = \lambda^{*-1}\tilde{\mathcal{U}}^\dagger \Pi_2. \tag{6.57}$$

Proof. (i) The identity

$$|\Pi_1 - \lambda\Pi_2|^* |\Pi_1| = |\Pi_2 - \lambda^*\Pi_1||\Pi_2|^* \qquad (6.58)$$

is verified, since the use of the equation

$$\Pi_2 = K^{-1}\Pi_2^{\dagger-1}\Pi_1^\dagger K\Pi_1 \qquad (6.59)$$

derived from (6.51) gives

$$
\begin{aligned}
|\Pi_2 - \lambda^*\Pi_1| &= |K^{-1}\Pi_2^{\dagger-1}\Pi_1^\dagger K - \lambda^*||\Pi_1| \\
&= |\Pi_2^{\dagger-1}\Pi_1^\dagger - \lambda^*||\Pi_1| \\
&= |\Pi_2|^{*-1}|\Pi_1 - \lambda\Pi_2|^*|\Pi_1|.
\end{aligned} \qquad (6.60)
$$

Accordingly, using (6.58) one obtains

$$|\Pi_1 - \lambda^{*-1}\Pi_2| = 0 \qquad (6.61)$$

for $\lambda\ (\neq 0)$ that satisfies

$$|\Pi_1 - \lambda\Pi_2| = 0. \qquad (6.62)$$

(ii) The assumption of (6.53) yields

$$(\Pi_1\mathcal{U}_a)^\dagger K(\Pi_1\mathcal{U}_b) = (\lambda_a\Pi_2\mathcal{U}_a)^\dagger K(\lambda_b\Pi_2\mathcal{U}_b), \qquad (6.63)$$

which leads to

$$\mathcal{U}_a^\dagger \tilde{K}\mathcal{U}_b = \lambda_a^*\lambda_b\mathcal{U}_a^\dagger\tilde{K}\mathcal{U}_b \qquad (6.64)$$

by the definition of $\tilde{K}$ of (6.52). Equation (6.64) means that when $\lambda_a^*\lambda_b \neq 1$

$$\mathcal{U}_a^\dagger \tilde{K}\mathcal{U}_b = 0 \qquad (6.65)$$

and moreover,

$$\mathcal{U}_b^\dagger \tilde{K}\mathcal{U}_a = (\mathcal{U}_a^\dagger \tilde{K}^\dagger\mathcal{U}_b)^\dagger = -(\mathcal{U}_a^\dagger\tilde{K}\mathcal{U}_b)^\dagger = 0. \qquad (6.66)$$

(iii) Multiplying $\tilde{K}\Pi_1^{-1}$ from the left side of (6.35), one obtains

$$\tilde{K}\Pi_1^{-1}\Pi_2\mathcal{U} = \lambda^{-1}\tilde{K}\mathcal{U}, \qquad (6.67)$$

and then finds the following for both sides of (6.67),

$$
\begin{aligned}
\tilde{K}\Pi_1^{-1}\Pi_2\mathcal{U} &= \Pi_1^\dagger K\Pi_2\mathcal{U} = \Pi_1^\dagger\tilde{\mathcal{U}} \\
\lambda^{-1}\tilde{K}\mathcal{U} &= \lambda^{-1}\Pi_2^\dagger K\Pi_2\mathcal{U} = \lambda^{-1}\Pi_2^\dagger\tilde{\mathcal{U}}.
\end{aligned} \qquad (6.68)
$$

Here (6.52) and (6.55) were employed.

$$\square$$

An immediate practical consequence of formula (i) is that the numbers of leftward decreasing evanescent waves ($|\lambda| > 1$) and rightward decreasing ones ($|\lambda| < 1$) are the same for a given energy E and lateral wave vector $k_\parallel$. On the other hand, for propagating Bloch waves ($|\lambda| = 1$), it follows from the conventional band theory that the numbers of left-propagating Bloch waves and right-propagating ones are also the same for a given E and $k_\parallel$; however, we give in Section 6.5 another rigorous verification of this postulate using the analyticity of the eigenvalues (see Theorems 6.5 and 6.6). Our procedure of the proof is advantageous, since it also leads to the peculiar properties of Green's functions that underlie the Green's function formalism discussed in Chapter 9 (see Theorems 9.1 and 9.2).

To date, computational methods relying on recursive calculation techniques, which are known as the transfer- (or propagation-)matrix method, have been developed for calculating scattering wave functions and generalized Bloch states. See, for instance, Marcus and Jepsen (1968), Appelbaum and Hamann (1972), Lee and Joannopoulos (1981a), Sautet and Joachim (1988), Takagaki and Ferry (1992), Sheng and Xia (1996), Mozos *et al.* (1997) and Wortmann *et al.* (2002). It is known however that when such recursive calculation techniques are employed as a means of numerically solving the Kohn–Sham equation, numerical errors frequently accumulate exponentially in the matrix elements of the recursively multiplicated transfer matrix [Holzwarth and Lee (1978); Wachutka (1986); Hummel and Bross (1998); Kobayashi (1999b); Choi and Ihm (1999)]. This difficulty is attributed to the appearance of exponentially growing and decaying evanescent waves [Wachutka (1986); Hirose and Tsukada (1994, 1995)]. In order to avoid the error accumulation, Wachutka (1986) proposed a way of improving the transfer-matrix methods by introducing a mathematical trick: The main difference from the original transfer-matrix methods is that the determination of wave functions ϕ is based on a boundary-value problem to pose a two-sided boundary condition for ϕ instead of an initial-value problem to pose a one-sided boundary condition for ϕ in the transfer-matrix approaches. Several methods including the present OBM formulation are basically in accord with Wachutka's concept for solving the Kohn–Sham equation as a boundary-value problem [Stiles and Hamann (1988); Hummel and Bross (1998); Kobayashi (1999b); Choi and Ihm (1999)]. However, we show below that Wachutka's proposal does not always work well. Indeed, the inherent instability in numerical calculations arises again when one attempts to include more intensively growing and decaying evanescent waves so as to describe the wave function more accurately.

In order to demonstrate the numerical accuracy of the generalized eigenvalue equation (6.35), we examine the eigenvalues calculated from (6.35).

We used the QZ algorithm with double precision, one of the standard solvers for a generalized eigenvalue problem. Figure 6.6 shows the product of all the moduli of the eigenvalues, $\prod_{n=1}^{2N} |\lambda_n|$. Taking into account the above-mentioned formula (i) of Theorem 6.3 which states that if λ_n is an eigenvalue of (6.35) then λ_n^{*-1} is necessarily an eigenvalue of (6.35), we see that an equation such that $\prod_{n=1}^{2N} |\lambda_n| = 1$ should be satisfied when λ_n's are the exact eigenvalues of (6.35). Nevertheless, the calculated value of $\prod_{n=1}^{2N} |\lambda_n|$ in Fig. 6.6 is found to be considerably far from 1 at a cutoff energy higher than ~ 10 Ry, corresponding to a grid spacing smaller than ~ 1.0 a.u. (The critical cutoff energy of the breaking depends on the computer and software employed, since numerical overflow and underflow are concerned.) This disagreement means that the eigenvalues, and therefore the eigenvectors as well, are smeared with numerical errors. We also ensured that at any grid spacing tested, the product of the moduli $|\lambda_n|$ satisfying $10^{-8} < |\lambda_n| < 10^8$ keeps equal to 1.

Consequently, we summarize the numerical accuracy of the generalized eigenvalue equation (6.35) as follows:

(i) Taking smaller spacings h_x and h_y, which correspond to the inclusion of transverse modes with larger kinetic energy, leads to the appearance of evanescent waves with exponential behavior that grow and decay more rapidly due to the moderation of the total energy.

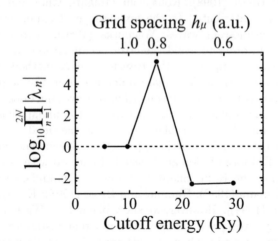

Fig. 6.6 Product of the moduli of the eigenvalues as a function of the cutoff energy or the grid spacing $h_\mu (= h_x = h_y = h_z)$ in the case of the Na jellium bulk ($r_s = 4.0$ a.u.). The seven-point finite-difference formula, i.e., $N_f = 3$ in (1.5), is used. The cutoff energy is expressed by $(\pi/h_\mu)^2$ (see Section 4.2).

When one chooses small h_x and h_y for implementing an accurate calculation, these rapidly growing and decaying evanescent waves give rise to numerical errors. This unfavorable situation can arise even when the calculation of the matrix elements of the generalized eigenvalue equation (6.35) is satisfactorily carried out.

(ii) Nevertheless, the propagating Bloch waves and gently growing or decaying evanescent waves can always be calculated with a high degree of accuracy.

The additional procedure for completely resolving the numerical problem associated with generalized Bloch states is shown in the next section.

6.4   Computational Scheme for Generalized Bloch States II. The Continued-Fraction Equations

We present an efficient and simple method for eliminating the above-mentioned numerical instability caused by the appearance of rapidly growing or decaying evanescent waves. This method is based on treating the ratios of the generalized Bloch states on the x–y planes at two successive z points, R^{ref} and R^{tra} [see Eq. (6.22) and the paragraph subsequent to it]. We will show that the Kohn–Sham equation of a crystalline bulk (see Fig. 6.4) is transformed into the continued-fraction equations with regard to R^{ref} and R^{tra}, whose solutions are self-consistently determined so as to agree with the periodicity of the z direction; thereby one can completely avoid the accumulation of numerical errors in evanescent waves. The *propagating* Bloch waves adopted as incident waves Φ^{in}, however, cannot be obtained from the continued-fraction equations, and therefore, for the implementation of the OBM formulas (6.25) and (6.26), we must use the continued-fraction method in combination with the method of the generalized eigenvalue equation (6.35) discussed in the preceding section. In addition, we note that numerical solutions of (6.35) with less accuracy are nevertheless of practical use, since they can be good initial estimates of R^{ref} and R^{tra} for self-consistent iterations of the continued-fraction equations.

Let us first deal with the ratios of the generalized Bloch states in the left electrode, R^{ref}. The matrix $Q^{ref}(z_k^M)$ $(= [\Phi_1^{ref}(z_k^M), ..., \Phi_N^{ref}(z_k^M)])$ given by (6.17), which gathers a set of N reflected waves that propagate and decay into the left electrode (see Fig 6.2), i.e., a set of generalized Bloch states with $|\lambda| \geq 1$, obeys the Kohn–Sham equation

$$-B(z_{k-1}^M)^\dagger Q^{ref}(z_{k-1}^M) + A(z_k^M)Q^{ref}(z_k^M) - B(z_k^M)Q^{ref}(z_{k+1}^M) = 0 \quad (6.69)$$

with $A(z_k^M)$ being $E - H(z_k^M; \mathbf{k}_{||})$. From (6.69), one can see that the ratio matrix R^{ref} of (6.22), that is,

$$R^{ref}(z_k^M) = Q^{ref}(z_{k-1}^M)Q^{ref}(z_k^M)^{-1} \tag{6.70}$$

satisfies a two-term recursive matrix equation

$$R^{ref}(z_{k+1}^M) = \left[A(z_k^M) - B(z_{k-1}^M)^{\dagger}R^{ref}(z_k^M)\right]^{-1}B(z_k^M). \tag{6.71}$$

Then the successive use of (6.71) leads to a continued-fraction representation, which connects $R^{ref}(z_1^{M+1})$ in the $(M+1)$-th unit cell to $R^{ref}(z_1^M)$ in the M-th unit cell as

$$R^{ref}(z_1^{M+1}) = \left[A(z_m^M) - B(z_{m-1}^M)^{\dagger}\left[A(z_{m-1}^M) - B(z_{m-2}^M)^{\dagger}\left[\cdots - B(z_2^M)^{\dagger}\right.\right.\right.$$
$$\times \left[A(z_2^M) - B(z_1^M)^{\dagger}\left[A(z_1^M) - B(z_m^{M-1})^{\dagger}R^{ref}(z_1^M)\right]^{-1}\right.$$
$$\times \left. B(z_1^M)\right]^{-1}B(z_2^M)\cdots\right]^{-1}B(z_{m-1}^M)\right]^{-1}B(z_m^M). \tag{6.72}$$

On the other hand, proceeding in the same direction, one finds that the ratio matrix R^{tra} of (6.22) in the right electrode defined by

$$R^{tra}(z_k^M) = Q^{tra}(z_k^M)Q^{tra}(z_{k-1}^M)^{-1} \tag{6.73}$$

obeys the matrix equation

$$R^{tra}(z_k^M) = \left[A(z_k^M) - B(z_k^M)R^{tra}(z_{k+1}^M)\right]^{-1}B(z_{k-1}^M)^{\dagger} \tag{6.74}$$

and the corresponding continued-fraction equation

$$R^{tra}(z_1^M) = \left[A(z_1^M) - B(z_1^M)\left[A(z_2^M) - B(z_2^M)\left[\cdots - B(z_{m-2}^M)\right.\right.\right.$$
$$\times \left[A(z_{m-1}^M) - B(z_{m-1}^M)\left[A(z_m^M) - B(z_m^M)R^{tra}(z_1^{M+1})\right]^{-1}\right.$$
$$\times \left. B(z_{m-1}^M)^{\dagger}\right]^{-1}B(z_{m-2}^M)^{\dagger}\cdots\right]^{-1}B(z_1^M)^{\dagger}\right]^{-1}B(z_m^{M-1})^{\dagger}. \tag{6.75}$$

In contrast to the case of the transfer matrix, (6.72) and (6.75) are stable recursion formulas without involving error accumulation. However, one sees that these formulas require a considerable amount of computation, since successive matrix inversions are demanded. In practice, instead of (6.72) and (6.75), one can employ more efficient formulas in terms of the Green's function matrix elements of (6.37). These formulas, equivalent to (6.72) and (6.75), are given in the following Theorem, which are derived based on the OBM scheme:

Theorem 6.4   *The ratio matrices R^{ref} and R^{tra} satisfy*

$$R^{ref}(z_1^{M+1}) = \mathcal{W}_{m,m}^M + \mathcal{W}_{m,1}^M \left[R^{ref}(z_1^M)^{-1} - \mathcal{W}_{1,1}^M \right]^{-1} \mathcal{W}_{1,m}^M$$

$$R^{tra}(z_1^M) = \mathcal{W}_{1,1}^M + \mathcal{W}_{1,m}^M \left[R^{tra}(z_1^{M+1})^{-1} - \mathcal{W}_{m,m}^M \right]^{-1} \mathcal{W}_{m,1}^M, \quad (6.76)$$

where $\mathcal{W}_{k,l}^M$ are defined by (6.37). The generalized Bloch conditions imposed on R^{ref} and R^{tra} read as

$$R^{ref}(z_k^{M+1}) = R^{ref}(z_k^M) \quad and \quad R^{tra}(z_k^{M+1}) = R^{tra}(z_k^M). \quad (6.77)$$

Proof.   It is sufficient to consider R^{ref} alone; the formulas concerning R^{tra} are similarly derived. From (6.17), (6.32) and (6.37), one finds that the sets of the generalized Bloch states $Q^{ref}(z_1^M)$ and $Q^{ref}(z_m^M)$ link together with the other sets $Q^{ref}(z_m^{M-1})$ and $Q^{ref}(z_1^{M+1})$ as

$$Q^{ref}(z_1^M) = \mathcal{W}_{1,1}^M Q^{ref}(z_m^{M-1}) + \mathcal{W}_{1,m}^M Q^{ref}(z_1^{M+1}) \quad (6.78)$$

and

$$Q^{ref}(z_m^M) = \mathcal{W}_{m,1}^M Q^{ref}(z_m^{M-1}) + \mathcal{W}_{m,m}^M Q^{ref}(z_1^{M+1}). \quad (6.79)$$

Multiplication of $Q^{ref}(z_m^{M-1})^{-1}$ from the right-hand side of (6.78) leads to

$$R^{ref}(z_1^M)^{-1} = \mathcal{W}_{1,1}^M + \mathcal{W}_{1,m}^M Q^{ref}(z_1^{M+1}) Q^{ref}(z_m^{M-1})^{-1}, \quad (6.80)$$

and similarly, (6.79) multiplied by $Q^{ref}(z_1^{M+1})^{-1}$ from the right-hand side takes the form

$$R^{ref}(z_1^{M+1}) = \mathcal{W}_{m,1}^M Q^{ref}(z_m^{M-1}) Q^{ref}(z_1^{M+1})^{-1} + \mathcal{W}_{m,m}^M. \quad (6.81)$$

Rewriting (6.80) as

$$Q^{ref}(z_m^{M-1}) Q^{ref}(z_1^{M+1})^{-1} = \left[R^{ref}(z_1^M)^{-1} - \mathcal{W}_{1,1}^M \right]^{-1} \mathcal{W}_{1,m}^M, \quad (6.82)$$

and inserting (6.82) into (6.81), one has the formula for R^{ref} in (6.76).

We next give the proof of (6.77) for R^{ref}. Taking into account the generalized Bloch condition of (6.33), one can write $Q^{ref}(z_k^{M+1})$ in the $(M+1)$-th unit cell as

$$Q^{ref}(z_k^{M+1}) = \left[\Phi_1^{ref}(z_k^{M+1}), \Phi_2^{ref}(z_k^{M+1}), ..., \Phi_N^{ref}(z_k^{M+1}) \right]$$

$$= \left[\lambda_1 \Phi_1^{ref}(z_k^M), \lambda_2 \Phi_2^{ref}(z_k^M), ..., \lambda_N \Phi_N^{ref}(z_k^M) \right], \quad (6.83)$$

where λ_i is the generalized Bloch factor for Φ_i^{ref}. Accordingly, the simple relationship between $R^{ref}(z_k^{M+1})$ and $R^{ref}(z_k^M)$,

$$R^{ref}(z_k^{M+1}) = Q^{ref}(z_{k-1}^{M+1})Q^{ref}(z_k^{M+1})^{-1}$$

$$= Q^{ref}(z_{k-1}^M) \begin{bmatrix} \lambda_1 & & 0 \\ & \ddots & \\ 0 & & \lambda_N \end{bmatrix} \begin{bmatrix} \lambda_1 & & 0 \\ & \ddots & \\ 0 & & \lambda_N \end{bmatrix}^{-1} Q^{ref}(z_k^M)^{-1}$$

$$= R^{ref}(z_k^M), \tag{6.84}$$

is established.   □

The solutions R^A ($A = ref$ and tra) of the continued-fraction equations (6.76) under the constraint of (6.77) are always determined in a self-consistent manner; in fact, one can numerically confirm that the correct solutions are obtained by several iterations of self-consistent cycles, when the crude solutions of (6.35) are adopted as initial estimates for R^A. Recall that the propagating Bloch waves and gently growing or decaying evanescent waves among the solutions of (6.35) are accurate, whereas the evanescent waves which rapidly grow or decay are smeared with serious numerical errors. It is then found that the preconditioning of the rapidly growing or decaying evanescent waves in the initial estimates such that they satisfy the conjugate relations (6.54) with the propagating Bloch waves and gentle evanescent waves can accelerate the convergence of self-consistent iterations.

6.5   Related Considerations

When defining the N-dimensional matrices Q^{ref} and Q^{tra} in the previous sections [see Eq. (6.17)], we have taken it for granted that the numbers of both the reflected waves Φ_n^{ref} and the transmitted waves Φ_n^{tra} are equally N for a given energy E and lateral Bloch wave vector $\mathbf{k}_{||}$; we remind readers that the waves Φ_n^{ref} (Φ_n^{tra}), which are composed of the leftward (rightward) propagating Bloch waves and decaying evanescent ones within the left (right) electrode (see Fig. 6.2), are the eigenvectors of the generalized eigenvalue equation (6.35), namely, the generalized Bloch states, belonging to the eigenvalues λ_n with $|\lambda_n| \geq 1$ ($|\lambda_n| \leq 1$). In this section, we show that this is the case for the numbers of Φ_n^{ref} and Φ_n^{tra}, and moreover give other miscellanies of importance related to the OBM formalism.

First we prove the following theorem [Lee and Joannopoulos (1981b)] that serves as a lemma of the proposition regarding the numbers of the generalized Bloch states Φ_n^{ref} and Φ_n^{tra} (Theorem 6.6) and in addition, yields asymptotic behavior of Green's functions (Theorem 9.1) and important relationships between the Green's functions and the ratio matrices R^{ref} and R^{tra} (Theorem 9.2), as will be discussed in Chapter 9.

Theorem 6.5 *Introducing an imaginary part into the energy E separates the $2N$ eigenvalues of the $2N$-dimensional generalized eigenvalue equation (6.35) equally into two groups of N eigenvalues with moduli greater and less than one, respectively. Namely, for nonreal Z ($= E + i\eta$), any eigenvalue $\lambda_n(Z, k_{||})$ satisfies $|\lambda_n(Z, k_{||})| \neq 1$, and the set $\{\lambda_n(Z, k_{||})\}$ can be divided into two halves with $|\lambda_n(Z, k_{||})| > 1$ and $|\lambda_n(Z, k_{||})| < 1$.*

Proof. We follow the prescription of Lee and Joannopoulos. Here $k_{||}$ is ignored for simplicity. It is first shown that the eigenvalue $\lambda_n(Z)$ must have the property $|\lambda_n(Z)| \neq 1$ if Z is off the real energy axis. Here, the $2N$-dimensional generalized eigenvalue equation for nonreal Z is

$$\Pi_1(Z) \begin{bmatrix} \Phi_n(z_m^{M-1}; Z) \\ \Phi_n(z_1^{M+1}; Z) \end{bmatrix} = \lambda_n(Z)\Pi_2(Z) \begin{bmatrix} \Phi_n(z_m^{M-1}; Z) \\ \Phi_n(z_1^{M+1}; Z) \end{bmatrix}, \qquad (6.85)$$

which is slightly different from (6.35), since $\mathcal{G}_T^M(z_k^M, z_l^M)$ in (6.37) is now the (k, l) block-matrix element of

$$\hat{\mathcal{G}}_T^M(Z) = \left[Z - \hat{H}_T^M \right]^{-1} \qquad (6.86)$$

instead of (6.31). Let us further consider the extended identity of (6.42), which is valid for a general complex number Z,

$$\left| Z - \hat{H}_T^M \right| \left| \Pi_1(Z) - \omega\Pi_2(Z) \right| = (-\omega)^N \left| Z - \hat{H}_{per}^M(\omega) \right|. \qquad (6.87)$$

Recalling that $\hat{H}_{per}^M(\omega = e^{ik_z L_z})$ with k_z of a real number is a Hermitian matrix [see Eq. (6.39)], one sees that its eigenvalues are real and then for nonreal Z,

$$\left| Z - \hat{H}_{per}^M(\omega = e^{ik_z L_z}) \right| \neq 0. \qquad (6.88)$$

This implies from (6.87) that for nonreal Z,

$$\left| \Pi_1(Z) - e^{ik_z L_z}\Pi_2(Z) \right| \neq 0, \qquad (6.89)$$

i.e., the eigenvalue $\lambda_n(Z)$ satisfying

$$\left| \Pi_1(Z) - \lambda_n(Z)\Pi_2(Z) \right| = 0 \qquad (6.90)$$

never has a modulus equal to one for Z off the real energy axis.

We proceed to show that there are equal numbers of $\lambda_n(Z)$ with modulus greater and less than one. From the argument using formula (i) of Theorem 6.3, we already know that if Z is on the real energy axis ($Z = E$), the set $\{\lambda_n(E)\}$ with $|\lambda_n(E)| \neq 1$ is divided *equally* into two groups with $|\lambda_n(E)| > 1$ and $|\lambda_n(E)| < 1$ (see the paragraph subsequent to the proof of Theorem 6.3). Furthermore, if E lies inside the energy region of band gaps (outside the bulk continuum), e.g., it becomes E_{bg} there, *all* $\lambda_n(E_{bg})$ have modulus greater or less than one. The proof of equal numbers of $\lambda_n(Z)$ with $|\lambda_n(Z)| > 1$ and $|\lambda_n(Z)| < 1$ for Z off the real energy axis is performed by *reductio ad absurdum*. Suppose that for some Z_0, there are not equal numbers of $\lambda_n(Z_0)$ with $|\lambda_n(Z_0)| > 1$ and $|\lambda_n(Z_0)| < 1$. Then, considering a path connecting the point Z_0 to E_{bg}, one must come across $\lambda_n(Z)$ with $|\lambda_n(Z)| = 1$ at a point $Z \neq$ real somewhere along this path, provided that $\lambda_n(Z)$ is continuous with respect to Z. This contradicts the fact that no $\{\lambda_n(Z)\}$ must have a modulus of one for nonreal Z.

The continuity of $\lambda_n(Z)$ is shown as follows. Consider the function

$$\mathcal{F}(\omega, Z) = \left| \Pi_1(Z)\Pi_2(Z)^{-1} - \omega \right|, \tag{6.91}$$

which is expressed in terms of $\{\lambda_n(Z)\}$ as

$$\mathcal{F}(\omega, Z) = \prod_{n=1}^{2N} (\lambda_n(Z) - \omega), \tag{6.92}$$

since $\{\lambda_n(Z)\}$ is also the set of eigenvalues of the standard eigenvalue equation transformed from (6.85), i.e.,

$$\Pi_1(Z)\Pi_2(Z)^{-1}\mathcal{U}'_n(Z) = \lambda_n(Z)\mathcal{U}'_n(Z), \tag{6.93}$$

where

$$\mathcal{U}'_n(Z) = \Pi_2(Z) \begin{bmatrix} \Phi_n(z_m^{M-1}; Z) \\ \Phi_n(z_1^{M+1}; Z) \end{bmatrix}. \tag{6.94}$$

From (6.87) we have another expression for $\mathcal{F}$ in the form of

$$\mathcal{F}(\omega, Z) = \left(\prod_{l=1}^{m} |B(z_l^M)| \right)^{-1} (-\omega)^N \left| Z - \hat{H}_{per}^M(\omega) \right|. \tag{6.95}$$

Here, we used the identity

$$\left| Z - \hat{H}_T^M \right| \left| \Pi_2(Z) \right| = \prod_{l=1}^{m} \left| B(z_l^M) \right|, \tag{6.96}$$

which can be obtained after some algebra including a limiting procedure as $\tau \equiv \omega^{-1} \to 0$ in (6.87). [Similarly, in the limit where $\omega \to 0$, (6.87) in turn reduces to another identity

$$\left| Z - \hat{H}_T^M \right| \left| \Pi_1(Z) \right| = \prod_{l=1}^{m} \left| B(z_l^M) \right|^*, \tag{6.97}$$

which is used in the proof of Theorem 9.1.] From (6.95) and the explicit representation of $\hat{H}_{per}^M(\omega)$, (6.39), it is obvious that $\mathcal{F}$ is an analytic *polynomial* in both ω and Z. Thus, one sees from (6.92) that the analyticity of $\mathcal{F}$ implies the continuity of $\lambda_n(Z)$. $\qquad\square$

Hereafter, let η in $Z = E + i\eta$ be a very small *positive* number. We now examine how to identify which eigenvalues $\{\lambda_n(E, \boldsymbol{k}_{||})\}$ with modulus equal to 1 analytically continue to $\{\lambda_n(E + i\eta, \boldsymbol{k}_{||})\}$ with modulus greater or less than 1. From what we learned above, no $\{\lambda_n(E + i\eta, \boldsymbol{k}_{||})\}$ can have modulus 1, and then the wave-vector component $k_z^{(n)}(E + i\eta, \boldsymbol{k}_{||})$ defined by

$$\lambda_n(E + i\eta, \boldsymbol{k}_{||}) = e^{ik_z^{(n)}(E+i\eta, \boldsymbol{k}_{||})L_z} \tag{6.98}$$

is complex. Expanding $k_z^{(n)}(E + i\eta, \boldsymbol{k}_{||})$ around E as

$$k_z^{(n)}(E + i\eta, \boldsymbol{k}_{||}) \simeq k_z^{(n)}(E, \boldsymbol{k}_{||}) + i\eta \frac{dk_z^{(n)}(E, \boldsymbol{k}_{||})}{dE}, \tag{6.99}$$

we have

$$\lambda_n(E + i\eta, \boldsymbol{k}_{||}) \simeq e^{ik_z^{(n)}(E, \boldsymbol{k}_{||})L_z} e^{-\eta L_z / v_n(E, \boldsymbol{k}_{||})}, \tag{6.100}$$

where v_n is the z component of the group velocity of the state (n), i.e.,

$$v_n(E, \boldsymbol{k}_{||}) = \frac{dE}{dk_z^{(n)}}. \tag{6.101}$$

Thus, the modulus of $\lambda_n(E + i\eta, \boldsymbol{k}_{||})$ is

$$\left| \lambda_n(E + i\eta, \boldsymbol{k}_{||}) \right| = \left| e^{ik_z^{(n)}(E+i\eta, \boldsymbol{k}_{||})L_z} \right|$$

$$\simeq e^{-\eta L_z / v_n(E, \boldsymbol{k}_{||})}. \tag{6.102}$$

From (6.102), one sees that the eigenvalues $\lambda_n(E, \boldsymbol{k}_{||})$ having modulus equal to 1 and associated with negative (positive) group velocity v_n, whose eigenvectors correspond to leftward (rightward) propagating Bloch waves, can analytically continue to $\lambda_n(E+i\eta, \boldsymbol{k}_{||})$ having modulus greater (less) than 1. That is to say, the eigenvectors belonging to the eigenvalues $\lambda_n(E + i\eta, \boldsymbol{k}_{||})$

with $\left|\lambda_n(E+i\eta,\boldsymbol{k}_{||})\right| > 1$ $\left(\left|\lambda_n(E+i\eta,\boldsymbol{k}_{||})\right| < 1\right)$ are the analytically con-
tinued vectors of the generalized Bloch states evaluated at the real energy
E, Φ_n^{ref} (Φ_n^{tra}), which are composed of the leftward (rightward) propagating
and decreasing waves. It is remarked that the group velocity v_n of (6.101)
only requires knowledge of $dE/dk_z^{(n)}$ at a purely *real* energy E; an explicit
procedure for calculating v_n will be given later [see Eq. (6.121)]. Thus,
combining the above argument with Theorem 6.5, we obtain the following
theorem.

Theorem 6.6   *There are always equal numbers of reflected waves Φ_n^{ref}
and transmitted ones Φ_n^{tra} for a given energy E and lateral Bloch wave vec-
tor $\boldsymbol{k}_{||}$. The respective sets of $\left\{\Phi_n^{ref}\right\}$ and $\left\{\Phi_n^{tra}\right\}$ consist of N eigenvectors
of the $2N$-dimensional generalized eigenvalue equation (6.35) belonging to
the eigenvalues $\lambda_n(E,\boldsymbol{k}_{||})$ with $\left|\lambda_n(E,\boldsymbol{k}_{||})\right| \geq 1$ and $\left|\lambda_n(E,\boldsymbol{k}_{||})\right| \leq 1$. A
propagating Bloch wave Φ_n specified by $\left|\lambda_n(E,\boldsymbol{k}_{||})\right| = 1$ is classified as a
reflected (transmitted) wave, according as its group velocity v_n is negative
(positive).*

For the sake of convenience, we summarize the relationships between
quantities on and off the real axis E. Let $\left\{\Phi_n(z_k^M;Z,\boldsymbol{k}_{||})\right\}$, $n = 1,2,...,N$,
be a set of the N eigenvectors of (6.85) with $\left|\lambda_n(Z,\boldsymbol{k}_{||})\right| > 1$, and
$Q(z_k^M;Z,\boldsymbol{k}_{||})$ be the N-dimensional matrix constructed by these eigenvectors, i.e.,

$$Q(z_k^M;Z,\boldsymbol{k}_{||}) = \left[\Phi_1(z_k^M;Z,\boldsymbol{k}_{||}),\Phi_2(z_k^M;Z,\boldsymbol{k}_{||}),...,\Phi_N(z_k^M;Z,\boldsymbol{k}_{||})\right].$$
(6.103)

These quantities should be compared with $\left\{\Phi_n^{ref}(z_k^M;E,\boldsymbol{k}_{||})\right\}$ and
$Q^{ref}(z_k^M;E,\boldsymbol{k}_{||})$ evaluated at a purely real energy E [see Eq. (6.17)]. More-
over, $R(z_k^M;Z,\boldsymbol{k}_{||})$ is defined by

$$R(z_k^M;Z,\boldsymbol{k}_{||}) = Q(z_{k-1}^M;Z,\boldsymbol{k}_{||})Q(z_k^M;Z,\boldsymbol{k}_{||})^{-1},$$
(6.104)

analogously to the definition of $R^{ref}(z_k^M;E,\boldsymbol{k}_{||})$, (6.70). It is then obvious
from Theorem 6.6 that

$$\lim_{\eta\to 0^+} \Phi_n(z_k^M;E+i\eta,\boldsymbol{k}_{||}) = \Phi_n^{ref}(z_k^M;E,\boldsymbol{k}_{||})$$

$$\lim_{\eta\to 0^+} Q(z_k^M;E+i\eta,\boldsymbol{k}_{||}) = Q^{ref}(z_k^M;E,\boldsymbol{k}_{||})$$

$$\lim_{\eta\to 0^+} R(z_k^M;E+i\eta,\boldsymbol{k}_{||}) = R^{ref}(z_k^M;E,\boldsymbol{k}_{||}).$$
(6.105)

On the other hand, proceeding along a similar line to the above argument, one obtains, for the quantities related to a set of eigenvectors

$\{\Phi_n(z_k^M; Z, \boldsymbol{k}_{||})\}$, $n = 1, 2, ..., N$, with $\left|\lambda_n(Z, \boldsymbol{k}_{||})\right| < 1$ [cf. Eqs. (6.17) and (6.73)],

$$\lim_{\eta \to 0^+} \Phi_n(z_k^M; E + i\eta, \boldsymbol{k}_{||}) = \Phi_n^{tra}(z_k^M; E, \boldsymbol{k}_{||})$$

$$\lim_{\eta \to 0^+} Q(z_k^M; E + i\eta, \boldsymbol{k}_{||}) = Q^{tra}(z_k^M; E, \boldsymbol{k}_{||})$$

$$\lim_{\eta \to 0^+} R(z_k^M; E + i\eta, \boldsymbol{k}_{||})^{-1} = R^{tra}(z_k^M; E, \boldsymbol{k}_{||}). \tag{6.106}$$

In the calculation of electronic transport, we are particularly concerned with the flux $\mathcal{I}(z)$, which is defined as the z component of the current through the unit-cell area $\mathcal{A}(z)$ on the x–y plane at z, i.e.,

$$\mathcal{I}(z) = \int_{\mathcal{A}(z)} d\boldsymbol{r}_{||} S_z(\boldsymbol{r}_{||}, z), \tag{6.107}$$

where S_z is the probability current density in the z direction,

$$S_z(\boldsymbol{r}_{||}, z) = -\frac{i}{2} \left[\phi^*(\boldsymbol{r}_{||}, z) \frac{\partial \phi(\boldsymbol{r}_{||}, z)}{\partial z} - \frac{\partial \phi^*(\boldsymbol{r}_{||}, z)}{\partial z} \phi(\boldsymbol{r}_{||}, z) \right], \tag{6.108}$$

and the wave function $\phi(\boldsymbol{r}_{||}, z)$ is the abbreviation of $\phi(\boldsymbol{r}_{||}, z; E, \boldsymbol{k}_{||})$, which is assumed to be normalized in the three-dimensional unit cell for convenience,

$$\int_{L_z} dz \int_{\mathcal{A}(z)} d\boldsymbol{r}_{||} \left|\phi(\boldsymbol{r}_{||}, z)\right|^2 = 1. \tag{6.109}$$

It is easily seen that $\mathcal{I}(z)$ is z-independent flux due to the equation of continuity, and $L_z\mathcal{I}$ represents the z component of the expectation value of the velocity operator $\hat{\boldsymbol{v}} = -i\nabla$ for a state ϕ.

It is straightforward to obtain the expression for the flux $\mathcal{I}(z)$ within the real-space finite-difference approach; with the following replacement for the first-order derivative,

$$\left. \frac{\partial \phi(\boldsymbol{r}_{||}, z)}{\partial z} \right|_{z=z_k^M} \to \frac{\phi(\boldsymbol{r}_{||}, z_{k+1}^M) - \phi(\boldsymbol{r}_{||}, z_k^M)}{h_z}, \tag{6.110}$$

(6.107) is rewritten in terms of the N-dimensional columnar vector $\Phi(z_k^M)$

as

$$\mathcal{I}(z_k^M) = -\frac{ih_x h_y}{2h_z}$$
$$\times \left[\Phi(z_k^M)^\dagger \left(\Phi(z_{k+1}^M) - \Phi(z_k^M) \right) - \left(\Phi(z_{k+1}^M)^\dagger - \Phi(z_k^M)^\dagger \right) \Phi(z_k^M) \right]$$
$$= ih_x h_y h_z \left[\Phi(z_k^M)^\dagger B(z_k^M) \Phi(z_{k+1}^M) - \Phi(z_{k+1}^M)^\dagger B(z_k^M)^\dagger \Phi(z_k^M) \right].$$
$$(6.111)$$

Here, we used (6.8) in the case of the central finite difference, i.e.,

$$B(z_k^M)^{(\dagger)} = -\frac{1}{2h_z^2} I. \tag{6.112}$$

More generally, for a state Φ not necessarily normalized in the unit cell, (6.111) reads as

$$\mathcal{I}(z_k^M) = i \left[\sum_{l=1}^{m} \Phi(z_l^M)^\dagger \Phi(z_l^M) \right]^{-1}$$
$$\times \left[\Phi(z_k^M)^\dagger B(z_k^M) \Phi(z_{k+1}^M) - \Phi(z_{k+1}^M)^\dagger B(z_k^M)^\dagger \Phi(z_k^M) \right]. \tag{6.113}$$

When a higher-order finite-difference formula is adopted, expression (6.113) for flux $\mathcal{I}$ holds with the simple replacement of z by ζ (see the next chapter).

Equation (6.113) satisfies the flux conservation and further has a proper relationship with the group velocity of (6.101), such as $v = L_z \mathcal{I}$, which is shown later in Theorem 6.7. The proof of the flux conservation is as follows. By using the Kohn–Sham equation

$$-B(z_k^M)^\dagger \Phi(z_k^M) + A(z_{k+1}^M) \Phi(z_{k+1}^M) - B(z_{k+1}^M) \Phi(z_{k+2}^M) = 0 \tag{6.114}$$

with $A(z_{k+1}^M)$ being $E - H(z_{k+1}^M; \mathbf{k}_{\|})$, we obtain, for the term in (6.113),

$$\Phi(z_k^M)^\dagger B(z_k^M) \Phi(z_{k+1}^M) - \Phi(z_{k+1}^M)^\dagger B(z_k^M)^\dagger \Phi(z_k^M)$$
$$= \Phi(z_k^M)^\dagger B(z_k^M) \Phi(z_{k+1}^M) - \Phi(z_{k+1}^M)^\dagger \left[A(z_{k+1}^M) \Phi(z_{k+1}^M) - B(z_{k+1}^M) \Phi(z_{k+2}^M) \right]$$
$$= \left[\Phi(z_k^M)^\dagger B(z_k^M) - \Phi(z_{k+1}^M)^\dagger A(z_{k+1}^M) \right] \Phi(z_{k+1}^M) + \Phi(z_{k+1}^M)^\dagger B(z_{k+1}^M) \Phi(z_{k+2}^M)$$
$$= \Phi(z_{k+1}^M)^\dagger B(z_{k+1}^M) \Phi(z_{k+2}^M) - \Phi(z_{k+2}^M)^\dagger B(z_{k+1}^M)^\dagger \Phi(z_{k+1}^M). \tag{6.115}$$

In the last step of (6.115), we employed the Hermitian-conjugate equation of (6.114), i.e.,

$$-\Phi(z_k^M)^\dagger B(z_k^M) + \Phi(z_{k+1}^M)^\dagger A(z_{k+1}^M) - \Phi(z_{k+2}^M)^\dagger B(z_{k+1}^M)^\dagger = 0. \tag{6.116}$$

Equation (6.115) implies that (6.113) is k-independent. $\qquad\square$

Comments concerning the 'off-diagonal flux' defined between two generalized Bloch states are added [Appelbaum and Blount (1973)]: Let Φ_n be a propagating wave. If $\Phi_{n'}$ is another propagating wave different from Φ_n or an evanescent wave, then the off-diagonal flux between Φ_n and $\Phi_{n'}$ vanishes, namely,

$$\Phi_n(z_k^M)^\dagger B(z_k^M)\Phi_{n'}(z_{k+1}^M) - \Phi_n(z_{k+1}^M)^\dagger B(z_k^M)^\dagger \Phi_{n'}(z_k^M) = 0. \quad (6.117)$$

In addition, let Φ_n be an evanescent wave with an eigenvalue λ_n of (6.35). If $\Phi_{n'}$ is a generalized Bloch state whose eigenvalue $\lambda_{n'}$ is not equal to λ_n^{*-1}, then the off-diagonal flux between these states vanishes as well. (Thus, the diagonal element of the flux for an evanescent wave is also zero.) We here prove (6.117) in both the cases. The left-hand side of (6.117) is denoted by $\mathcal{I}_{nn'}(z_k^M)$. The flux conservation is found to hold in the present cases, and then we have

$$\mathcal{I}_{nn'}(z_k^{M+1}) = \mathcal{I}_{nn'}(z_k^M). \quad (6.118)$$

Using the (generalized) Bloch conditions

$$\Phi_n(z_k^{M+1}) = \lambda_n \Phi_n(z_k^M) \quad \text{and} \quad \Phi_{n'}(z_k^{M+1}) = \lambda_{n'}\Phi_{n'}(z_k^M), \quad (6.119)$$

where $\lambda_n = e^{ik_n L_z}$ and $\lambda_{n'} = e^{ik_{n'} L_z}$ (for evanescent waves, replace k_n and $k_{n'}$ by $i\kappa_n$ and $i\kappa_{n'}$, respectively), and noting the periodicity of $\{B(z_k^M)\}$, i.e., $B(z_k^{M+1}) = B(z_k^M)$, we obtain

$$\begin{aligned}
\mathcal{I}_{nn'}(z_k^{M+1}) &\equiv \Phi_n(z_k^{M+1})^\dagger B(z_k^{M+1})\Phi_{n'}(z_{k+1}^{M+1}) \\
&\quad - \Phi_n(z_{k+1}^{M+1})^\dagger B(z_k^{M+1})^\dagger \Phi_{n'}(z_k^{M+1}) \\
&= \lambda_n^* \lambda_{n'} \mathcal{I}_{nn'}(z_k^M). \quad (6.120)
\end{aligned}$$

The assumption reads as $\lambda_n^* \lambda_{n'} \neq 1$. This establishes $\mathcal{I}_{nn'}(z_k^M) = 0$.

□

Theorem 6.7 *The z component of the group velocity, v ($\equiv dE/dk_z$) of (6.101), has an explicit representation in the form of*

$$v(E, \boldsymbol{k}_{\|}) = L_z \mathcal{I}$$

$$= iL_z \left[\sum_{l=1}^m \Phi(z_l^M)^\dagger \Phi(z_l^M) \right]^{-1}$$

$$\times \left[\Phi(z_k^M)^\dagger B(z_k^M)\Phi(z_{k+1}^M) - \Phi(z_{k+1}^M)^\dagger B(z_k^M)^\dagger \Phi(z_k^M) \right]. \quad (6.121)$$

Proof. Differentiating the eigenvalue equation of (6.38), i.e.,

$$\hat{H}_{per}(\omega = e^{ik_z L_z}, \mathbf{k}_{||})\hat{\Phi} = E\hat{\Phi} \tag{6.122}$$

with respect to k_z and multiplying $\hat{\Phi}^\dagger$ from the left-hand side of the resultant equation yield

$$\hat{\Phi}^\dagger \frac{d\hat{H}_{per}}{dk_z}\hat{\Phi} = \frac{dE}{dk_z}\hat{\Phi}^\dagger\hat{\Phi}, \tag{6.123}$$

where the eigenvector $\hat{\Phi}$ is an mN-dimensional vector

$$\hat{\Phi} = \begin{bmatrix} \Phi(z_1^M) \\ \Phi(z_2^M) \\ \vdots \\ \Phi(z_m^M) \end{bmatrix}. \tag{6.124}$$

[Note that an equation like (6.123) is referred to as that of the Hellmann–Feynman theorem.] From the representation of $\hat{H}_{per}$, (6.39), one finds that

$$\frac{d\hat{H}_{per}}{dk_z} = \begin{bmatrix} 0 & 0\cdots 0 & -iL_z\omega^{-1}B(z_m^M)^\dagger \\ 0 & 0\cdots 0 & 0 \\ \vdots & \vdots \quad \vdots & \vdots \\ iL_z\omega B(z_m^M) & 0\cdots 0 & 0 \end{bmatrix} \tag{6.125}$$

and then

$$\hat{\Phi}^\dagger \frac{d\hat{H}_{per}}{dk_z}\hat{\Phi} = iL_z\omega\Phi(z_m^M)^\dagger B(z_m^M)\Phi(z_1^M)$$

$$-iL_z\omega^{-1}\Phi(z_1^M)^\dagger B(z_m^M)^\dagger\Phi(z_m^M)$$

$$= iL_z\Phi(z_m^M)^\dagger B(z_m^M)\Phi(z_1^{M+1})$$

$$-iL_z\Phi(z_1^{M+1})^\dagger B(z_m^M)^\dagger\Phi(z_m^M). \tag{6.126}$$

Here, the last step follows from the Bloch condition

$$\Phi(z_1^{M+1}) = \omega\Phi(z_1^M). \tag{6.127}$$

Thus, from (6.123) and (6.126), one can obtain (6.121) at once, using the flux conservation and the expression of the normalization of $\hat{\Phi}$. $\quad\square$

Finally, we give the procedure for calculating the conductance in the OBM method. When only the Γ point, $\mathbf{k}_{||} = (0,0)$, is taken into account in the transport calculation, the conductance G at the zero bias limit is described by the following formula of Landauer–Büttiker type [see Eq. (6.6)]:

$$G = \frac{2e^2}{h} \sum_{i,j} |t_{ij}|^2 \frac{v_i'}{v_j}, \tag{6.128}$$

where v_j and v_i' are the group velocities in the z direction of the j-th incident Bloch wave Φ_j^{in} and the i-th transmitted Bloch wave Φ_i^{tra}, respectively, and t_{ij} represents the corresponding transmission coefficient which are evaluated at the Fermi level. Here the left and right electrodes are assumed to be two different materials in general. In the OBM formalism, the scattering wave function

$$\Psi_j(z_k) = \sum_i t_{ij} \Phi_i^{tra}(z_k), \tag{6.129}$$

where $k = m + 2$ or $m + 1$, is directly determined by solving (6.25) or (6.26). Since the transition region is chosen to be large enough that the potential is effectively screened, $\Psi_j(z_k)$ in the right-electrode region is actually composed of propagating Bloch waves alone with negligible contribution of evanescent waves. Hence, the transmission coefficients t_{ij} can be determined from the following equation for given values of $\Psi_j(z_k)$ and transmitted *propagating* Bloch waves $\Phi_i^{tra}(z_k)$ ($k = m + 2$ or $m + 1$):

$$\begin{bmatrix} \Phi_1^{tra\dagger}\Phi_1^{tra} & \cdots & \Phi_1^{tra\dagger}\Phi_{N_{tp}}^{tra} \\ \vdots & & \vdots \\ \Phi_{N_{tp}}^{tra\dagger}\Phi_1^{tra} & \cdots & \Phi_{N_{tp}}^{tra\dagger}\Phi_{N_{tp}}^{tra} \end{bmatrix} \begin{bmatrix} t_{1,j} \\ \vdots \\ t_{N_{tp},j} \end{bmatrix} = \begin{bmatrix} \Phi_1^{tra\dagger}\Psi_j \\ \vdots \\ \Phi_{N_{tp}}^{tra\dagger}\Psi_j \end{bmatrix}, \tag{6.130}$$

where N_{tp} stands for the number of transmitted propagating Bloch waves. The group velocities in (6.128) are evaluated from (6.121). It is noted that in the case of including the contributions from the first Brillouin zone of $k_{||}$ ($|k_x| \leq \pi/L_x, |k_y| \leq \pi/L_y$), formula (6.128) is extended to

$$G = \frac{2e^2}{h} \frac{L_x L_y}{(2\pi)^2} \int_{1BZ} dk_{||} |t_{i,j}|^2 \frac{v_i'}{v_j}, \tag{6.131}$$

since

$$\int_{1BZ} dk_{||} f(k_{||}) \simeq \frac{2\pi}{L_x} \times \frac{2\pi}{L_y} \times f(k_{||} = 0), \tag{6.132}$$

which guarantees that (6.128) is a good approximation of (6.131) in a large supercell.

6.6   Application: Sodium Nanowires

In this section, an application of the OBM method is presented to explain conduction properties of the closed packed structure and single-row wires consisting of sodium atoms suspended between genuine crystalline electrodes. The sodium wire is one of the most frequently examined wires in both experiments [Krans *et al.* (1995, 1996)] and theoretical calculations [Lang (1997a, 1997b); Barnett and Landman (1997); Nakamura *et al.* (1999); Kobayashi (1999a); Kobayashi *et al.* (2000); Sim *et al.* (2001, 2002); Havu *et al.* (2002); Tsukamoto and Hirose (2002)], since the conductance quantization emerges best in the wires consisting of monovalent atoms.

The computational conditions are as follows: a grid spacing is chosen to be 0.81 a.u., which corresponds to a cutoff energy of 15 Ry in the plane-wave approach. We employ the central finite-difference formula, i.e., $N_f = 1$ in (1.5), for the derivative arising from the kinetic-energy operator, the local pseudopotential [Bachelet *et al.* (1982)] for Coulomb potential between electrons and nucleus, and the local density approximation [Perdew and Zunger (1981)] for the exchange-correlation interaction between electrons.

Figure 6.7 illustrates an example of the calculation models, in which a single-row sodium wire is sandwiched by the square bases made of sodium atoms and all of them are suspended between the Na(001) crystalline elec-

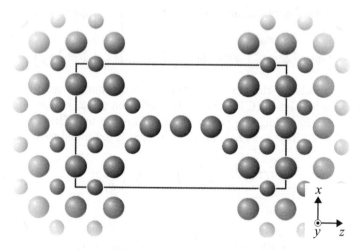

Fig. 6.7   Atomic configuration for the $N_{atom} = 3$ single-row sodium wire suspended between semi-infinite crystalline electrodes. The rectangle represents the supercell used to evaluate the optimized atomic configuration and electronic structure.

trodes. The distance between the electrode surface and the basis, as well as that between the basis and the edge atom of the wire, is set to be $a_0/2$, where a_0 (= 8.1 a.u.) is the lattice constant of the sodium crystal. The number of atoms composing the wire, N_{atom}, is varied between one and five, and the electrode spacing is set to be $2a_0 + (N_{atom} - 1) \times \sqrt{3}/2a_0$. This structure forms the scattering region in the conductance calculations. Periodic boundary conditions are imposed on the supercell in the x and y directions perpendicular to the wire axis. The side lengths of the supercell in the x and y directions are $3a_0$. We first determine the relaxed atomic structure and the electronic structure of the scattering region by employing a conventional supercell technique; the supercell indicated by the rectangle in Fig. 6.7 is used. During the structural optimization, the atoms inside the wire are optimized along the z direction while the other atoms are kept frozen. Then, we evaluate electric conductances of the wires at the Γ point, $k_{||} = (0,0)$, by using the OBM method.

Figure 6.8 shows the calculated conductances as a function of the number N_{atom} of atoms in the wires. The conductances are ~ 3 G_0 ($G_0 = 2e^2/h$) for the closed packed structure ($N_{atom} = 1$), and ~ 1 G_0 for the single-row wires ($N_{atom} = 2 - 5$). In addition, we observed the even-odd oscillation of the conductance in which the conductances for the even atom numbers ($N_{atom} = 2$ and 4) are slightly lower than 1 G_0 whereas those for the odd atom numbers ($N_{atom} = 3$ and 5) are close

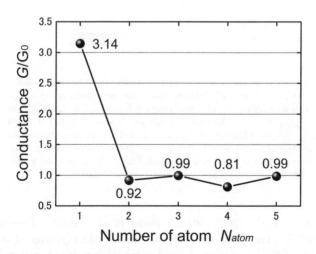

Fig. 6.8 Conductances of closed packed structure ($N_{atom} = 1$) and single-row sodium wires ($N_{atom} = 2 - 5$). Reprinted with permission from Egami *et al.* (2004). © 2004 The Japan Institute of Metals.

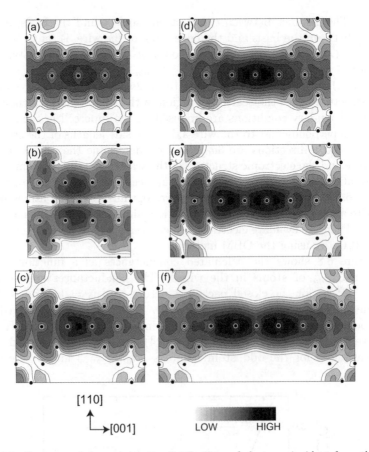

[110]
→[001]

LOW   HIGH

Fig. 6.9 Contours of channel density distributions of electrons incident from the left electrode at the Fermi level. The planes shown are perpendicular to the [110] direction and include the wire axis. (a) The first channel for $N_{atom} = 1$, (b) the second (third) channel for $N_{atom} = 1$, (c) the first channel for $N_{atom} = 2$, (d) the first channel for $N_{atom} = 3$, (e) the first channel for $N_{atom} = 4$, and (f) the first channel for $N_{atom} = 5$. Each contour represents twice or half density of the adjacent contour lines. The lowest contour represents 1.31×10^{-9} electron/bohr^3/eV. Reprinted with permission from Egami *et al.* (2004). © 2004 The Japan Institute of Metals.

to 1 G$_0$ [Sim *et al.* (2001, 2002); Havu *et al.* (2002); Tsukamoto and Hirose (2002)]. The previous experimental studies [Krans *et al.* (1995); Krans *et al.* (1996)] on electron conduction properties of sodium wires observed that there are marked peaks at 1, 3, 5, and 6 G$_0$ in the histogram of conductance and no peaks at 2, 4, and 7 G$_0$. The present calculation results that the single-row sodium wire exhibits the quantized conductance

Table 6.1   Channel transmissions at the Fermi level for $N_{atom} = 1 - 5$. Data taken from Egami *et al.* (2004)

N_{atom}	1	2	3	4	5
1st ch.	1.00	0.869	0.994	0.814	0.986
2nd ch.	0.83	0.024	0.000	0.000	0.000
3rd ch.	0.83	0.024	0.000	0.000	0.000
4th ch.	0.27	0.000	0.000	0.000	0.000
5th ch.	0.15	0.000	0.000	0.000	0.000

of ~ 1 G_0 and ~ 3 G_0, but does not manifest a conductance of 2 G_0 agree with the results of the previous experiments.

For more detailed interpretation of electron conduction properties in the sodium wire, we next analyze individual conduction channels. The eigen-channels are investigated by diagonalizing the Hermitian matrix $(\mathbf{T}^\dagger\mathbf{T})$, where $\mathbf{T}$ is a transmission-coefficient matrix whose matrix elements are represented by

$$(\mathbf{T})_{ij} = t_{ij}\sqrt{\frac{v_i'}{v_j}} \tag{6.133}$$

in terms of the factors appearing in the Landauer–Büttiker formula (6.128) [Kobayashi *et al.* (2000)]. Table 6.1 collects the transmissions of the incident electrons for the respective conduction channels. The channel density distributions for electrons incident from the left electrode at the Fermi level are illustrated in Fig. 6.9, where the planes shown are perpendicular to the [110] direction and include the wire axis. In the case of the closed packed structure ($N_{atom} = 1$), the electrode spacing is so short that there are three channels mainly contributing to electron conduction: the first channel conducts electrons through the wire itself, while the second and third channels conduct them directly from the left base to the right base, which yields higher conduction of the closed packed structure against that of single-row wires. In the cases of $N_{atom} = 2 - 5$, since the electrode spacing is sufficiently large, only the first channel, which transmits electrons along the wire itself, opens.

Moreover, in Fig. 6.9, one can see that the bunches of high charge densities whose lengths are twice the interatomic distance in the wire are preferentially formed in the longer wires. Since the Fermi wavelength in the sodium crystal is ~ 13.1 a.u., which corresponds to twice the inter-atomic distance in the wire, the even-odd behavior of the conductance may be attributed to the relationship between the wire length and the Fermi wavelength. That is, the local structure of the scattering region plays an important role in electron conduction. In addition, it is well known that the density wave with the period of the two atoms emerges in an elongated infinite single-row sodium wire [Fröhlich (1954); Peierls (1955); Tsukamoto *et al.* (2001); Ono *et al.* (2003b)], and therefore, the characteristic of the channel wave function can be affected by the density wave. Further details of the simulations are presented in Egami *et al.* (2004) and elsewhere.

Chapter 7

Inclusion of Norm-Conserving Pseudopotentials

Tomoya Ono and Kikuji Hirose

7.1 Treatment of Nonlocal Pseudopotentials

In this section, we extend the overbridging boundary-matching (OBM) scheme to the case that includes the norm-conserving pseudopotentials in order to calculate the conduction properties of the nanostructures made of multivalent atoms with s-, p-, and d-valence orbitals. When we treat a set of values of the wave function on the x–y plane as a columnar vector, the nonlocal parts of the norm-conserving pseudopotentials are represented by $(N_{in} \times N_{xy})$-dimensional block matrices in the Kohn–Sham Hamiltonian as shown, for example, by the region depicted with oblique lines in Fig. 7.1. Here, N_{in} is the number of the x–y planes crossing the inner-shell region of the pseudopotential and N_{xy} $(= N_x \times N_y)$ is the number of grid points on the x–y plane. Thus, in the case of including the nonlocal parts, the Hamiltonian is no longer denoted by a block-tridiagonal matrix, as long as the central finite-difference formula is adopted for the kinetic-energy operator. We then introduce an appropriate higher-order finite-difference formula so that the off-diagonal elements of the finite-difference formula can entirely cover the block matrices of the nonlocal parts of the pseudopotentials in the semi-infinite *left*- and *right-electrode* regions; for this purpose, the order of the finite-difference approximation N_f [see Eq. (1.5)] is chosen to be greater than half the N_{in} of atoms in the electrodes. As a consequence, the Hamiltonian is again described in a block-tridiagonal form in the electrode regions, in which block-matrix elements are N $(= N_f \times N_{xy})$ dimensional. It is noted that the block tridiagonality of the Hamiltonian is not necessarily needed in the transition region to realize wave-function matching based on the OBM scheme. The areas shaded dark grey and light grey

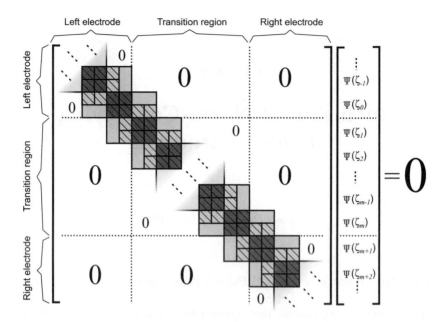

Fig. 7.1   Schematic representation of the discretized Kohn–Sham equation in the case of including the nonlocal pseudopotentials. Small squares denote N_{xy} ($= N_x \times N_y$)-dimensional block-matrix elements.

in Fig. 7.1 correspond to diagonal block-matrix elements $H(\zeta_k; \mathbf{k}_{||})$ and off-diagonal block-matrix elements $B(\zeta_k)^{(\dagger)}$ [cf. Eq. (6.10)], respectively, where ζ_k stands for the group index of z coordinates within the closed interval $[z_{(k-1)N_f+1}, z_{kN_f}]$, as illustrated in Fig. 7.2. A set of values of the wave function on the x–y planes within the interval $[z_{(k-1)N_f+1}, z_{kN_f}]$ is treated as an N-dimensional columnar vector $\Psi(\zeta_k)$ (see Fig. 7.1). This way, the Hamiltonian in the electrode regions is rewritten by a block-tridiagonal matrix consisting of N-dimensional block matrices.

We summarize the formulas for calculating the scattering wave function of the whole system in the presence of the nonlocal pseudopotentials, which are a straightforward extension of the OBM formulas given in the preceding chapter. Firstly, the $2N$ generalized Bloch states Φ inside the electrodes are computed as the eigenvectors of the $2N$-dimensional generalized eigenvalue equation [cf. Eq. (6.35) in Theorem 6.1] given by

$$\Pi_1(E, \mathbf{k}_{||}) \begin{bmatrix} \Phi(\zeta_m^{M-1}) \\ \Phi(\zeta_1^{M+1}) \end{bmatrix} = \lambda \Pi_2(E, \mathbf{k}_{||}) \begin{bmatrix} \Phi(\zeta_m^{M-1}) \\ \Phi(\zeta_1^{M+1}) \end{bmatrix}. \tag{7.1}$$

Subsequently, the N-dimensional ratio matrices of R^{ref} and R^{tra} defined by

$$R^{ref}(\zeta_k^M) = Q^{ref}(\zeta_{k-1}^M)Q^{ref}(\zeta_k^M)^{-1}$$
$$R^{tra}(\zeta_k^M) = Q^{tra}(\zeta_k^M)Q^{tra}(\zeta_{k-1}^M)^{-1} \qquad (7.2)$$

are obtained by solving the continued-fraction equations [cf. Eq. (6.76) in Theorem 6.4]

$$R^{ref}(\zeta_1^{M+1}) = \mathcal{W}_{m,m}^M + \mathcal{W}_{m,1}^M \left[R^{ref}(\zeta_1^M)^{-1} - \mathcal{W}_{1,1}^M \right]^{-1} \mathcal{W}_{1,m}^M$$
$$R^{tra}(\zeta_1^M) = \mathcal{W}_{1,1}^M + \mathcal{W}_{1,m}^M \left[R^{tra}(\zeta_1^{M+1})^{-1} - \mathcal{W}_{m,m}^M \right]^{-1} \mathcal{W}_{m,1}^M \quad (7.3)$$

under the generalized Bloch conditions

$$R^{ref}(\zeta_k^{M+1}) = R^{ref}(\zeta_k^M) \quad \text{and} \quad R^{tra}(\zeta_k^{M+1}) = R^{tra}(\zeta_k^M). \qquad (7.4)$$

Here, Π and $\mathcal{W}^M$ are the matrices defined in the same manner as (6.36) and (6.37), respectively, and $Q^{ref}(Q^{tra})$ is the matrix like (6.17), i.e., the following N-dimensional matrix constructed by the N reflected waves Φ^{ref} (transmitted waves Φ^{tra}) that propagate and decay into the left (right) electrodes, as seen in Fig. 7.2, and are, at the same time, the N eigenvectors of (7.1) with $|\lambda| \geq 1$ ($|\lambda| \leq 1$):

$$Q^A(\zeta_k^M) = \left[\Phi_1^A(\zeta_k^M), \Phi_2^A(\zeta_k^M), \dots, \Phi_N^A(\zeta_k^M) \right] \quad (A = ref \text{ and } tra). \quad (7.5)$$

It is emphasized again that in some cases, among the eigenvectors of (7.1), rapidly growing and decaying evanescent waves are smeared with numerical errors, and that self-consistent calculations employing the continued-

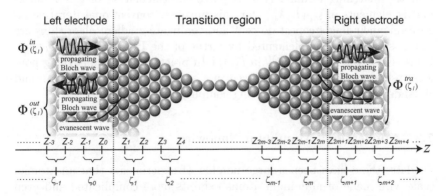

Fig. 7.2 Sketch of the relationship between z and ζ in the computational model in the case of $N_f = 2$. The meanings of the other parameters are the same as those in Fig. 6.2.

fraction equations (7.3) under the restrictions (7.4) can completely eliminate this numerical problem to yield accurate ratio matrices R^{ref} and R^{tra} (see Sections 6.3 and 6.4). Finally, using the ratio matrices R^{ref} and R^{tra} thus determined, the scattering wave function of the whole system, Ψ, for an incident wave Φ^{in} propagating from deep inside the left electrode is calculated as a solution of the $2N$-dimensional simultaneous equations of the wave-function matching formula [cf. Eqs. (6.26) and (6.21)]

$$\begin{bmatrix} \mathcal{W}_{0,0}R^{ref}(\zeta_0) - I & \mathcal{W}_{0,m+1}R^{tra}(\zeta_{m+2}) \\ \mathcal{W}_{m+1,0}R^{ref}(\zeta_0) & \mathcal{W}_{m+1,m+1}R^{tra}(\zeta_{m+2}) - I \end{bmatrix} \begin{bmatrix} \Psi(\zeta_0) \\ \Psi(\zeta_{m+1}) \end{bmatrix}$$

$$= - \begin{bmatrix} \mathcal{W}_{0,0} \\ \mathcal{W}_{m+1,0} \end{bmatrix} (\Phi^{in}(\zeta_{-1}) - R^{ref}(\zeta_0)\Phi^{in}(\zeta_0)), \qquad (7.6)$$

where

$$\mathcal{W}_{0,0} = \mathcal{G}_T(\zeta_0, \zeta_0)B(\zeta_{-1})^\dagger$$
$$\mathcal{W}_{0,m+1} = \mathcal{G}_T(\zeta_0, \zeta_{m+1})B(\zeta_{m+1})$$
$$\mathcal{W}_{m+1,0} = \mathcal{G}_T(\zeta_{m+1}, \zeta_0)B(\zeta_{-1})^\dagger \qquad (7.7)$$
$$\mathcal{W}_{m+1,m+1} = \mathcal{G}_T(\zeta_{m+1}, \zeta_{m+1})B(\zeta_{m+1})$$

and $\mathcal{G}_T(\zeta_k, \zeta_l)$ is the N-dimensional (k, l) block-matrix element of the Green's function matrix

$$\hat{\mathcal{G}}_T(E, \boldsymbol{k}_{||}) = \left[E - \hat{H}_T(\boldsymbol{k}_{||}) \right]^{-1} \qquad (7.8)$$

with $\hat{H}_T(\boldsymbol{k}_{||})$ being the $(m+2)N$-dimensional truncated Hamiltonian relative to z coordinates within the closed interval $[\zeta_0, \zeta_{m+1}]$. We stress again that the matching formula (7.6) requires the calculation of a few block-matrix elements of $\hat{\mathcal{G}}_T(E, \boldsymbol{k}_{||})$; for this purpose, conventional iterative algorithms including steepest-descent and conjugate-gradient algorithms are the most efficiently implemented by virtue of the Hermiticity of $\hat{H}_T(\boldsymbol{k}_{||})$. [This is also the case for $\mathcal{W}_{k,l}^M$ in (7.3).] In practice, norm-conserving nonlocal pseudopotentials are employed in a separable form of Kleinman and Bylander [see Eq. (1.23)] for saving computational costs.

7.2 Application: Aluminum Nanowire

To date, there have been a large number of experiments that an atomic-scale wire made of aluminum atoms exhibits more complicated and even more interesting features, that is, the plateau of conductance steps is not flat but has a positive slope before the wire breaks [Krans *et al.* (1993); Scheer *et al.* (1997); Cuevas *et al.* (1998); Mizobata *et al.* (2003a, 2003b)].

This somewhat counterintuitive result implies that conductance increases even when the electrodes are pulled apart.

In this section, we demonstrate first-principles calculations to elucidate the close relationship between the relaxed geometrical structure and electronic conductance of a three-aluminum-atom wire during its elongation using a model in which the wire is suspended between two semi-infinite Al(001) crystalline electrodes. The electron transmission is calculated by the OBM procedure discussed above. In order to include the contribution from the both s- and p-valence electrons correctly, the electron-ion interaction is described by the norm-conserving pseudopotentials [Kobayashi (1999a)] of Troullier and Martins (1991). The nine-point finite-difference case ($N_f = 4$) is adopted for the derivative arising from the kinetic-energy operator in the Kohn–Sham equation. Exchange-correlation effects are treated by the local-density approximation [Perdew and Zunger (1981)] in the density-functional theory. Although the atomic configuration just before breaking is not yet clear in experiments, we chose a three-atom wire the most appropriate configuration in view of structural stability and conduction property according to the background of previous experiments and first-principles calculations [Lang (1995); Kobayashi *et al.* (2000); Yanson (2001); Palacios *et al.* (2002); Lee *et al.* (2003)].

First, we determine the relaxed atomic structure of the three-aluminum-atom wire suspended between the Al(001) electrodes during its elongation. A cutoff energy is set at 25 Ry, which corresponds to a grid spacing of 0.63 a.u., and further a higher cutoff energy of 224 Ry is taken in the vicinity of nuclei with the augmentation of double-grid points. For structural optimization, a conventional supercell, which is denoted by the rectangle enclosed by broken lines in Fig. 7.3, is employed. The size of the supercell is chosen as $L_x = L_y = 15.12$ a.u. and $L_z = 22.68$ a.u.$+2d$. Here, the z axis is taken along the wire axis, the x and y axes are perpendicular to the wire, L_x, L_y and L_z are the lengths of the supercell in the x, y and z directions, respectively, and d is the average value for the projections of interatomic distances between adjacent atoms of the wire onto the z component. The wire is attached at both its ends to the square basis made of aluminum atoms with side lengths of $a_0/\sqrt{2}$, modeled after the [001] aluminum strands, and all of these components are sandwiched between the electrodes consisting of three atomic layers of the Al(001) surface. We take the distance between the electrode surface and the basis, as well as that between the basis and the edge atom of the wire, to be $0.5a_0$, where $a_0 (= 7.56$ a.u.$)$ is the lattice constant of the aluminum crystal. The center atom of the wire is relaxed on the (110) plane while the other atoms are fixed during the first-principles structural optimization. The wire exhibits a bent structure below an electrode spacing L_{es} of 26.46 a.u. Elongated up

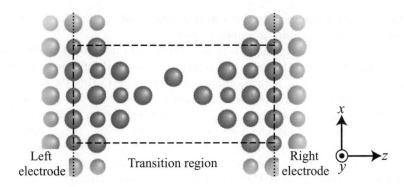

Fig. 7.3   Schematic representation of the model. The rectangle enclosed by broken lines represents the supercell used to evaluate the optimized atomic configuration and electronic structure.

to $L_{es} = 27.09$ a.u., the wire deforms from a bent structure to a straight one, and finally distorts at $L_{es} = 28.35$ a.u. We also implemented structural optimization using a large supercell of $L_x = L_y = 21.40$ a.u. and $L_z = 22.68$ a.u.$+2d$, and found no qualitative difference between these results and those in the case of a supercell of $L_x = L_y = 15.12$ a.u. adopted in the text. This implies that a supercell length of 15.12 a.u. is adequate to eliminate the unexpected interaction from the wires in neighboring cells.

We next explore the electron transmission of the wire at the zero-bias limit with the Landauer–Büttiker formula $G = \mathrm{tr}(\mathbf{T}^\dagger\mathbf{T})G_0$, where $\mathbf{T}$ is a transmission matrix [see Eqs. (6.128) and (6.133)]. In order to obtain the effective potential required in the OBM formalism, the electronic structure of the transition region which is composed of wire, bases, and a couple of atomic layers of electrodes (see Fig. 7.3) as well as the electronic structure of the aluminum crystalline electrode is calculated self-consistently employing a conventional supercell technique with a three-dimensional periodic boundary.

Figure 7.4 shows channel transmission at the Fermi level for the optimized wire as a function of L_{es}. We find certain characteristics that are common to the unbroken three-aluminum-atom wire: throughout all the ranges of electrode spacing below $L_{es} = 28.35$ a.u., the first channel, which has the characters of s and p_z for a straight wire, is widely open, while the second and third ones, which have the characters of degenerate p_x and p_y for the straight wire, have a small transmission of $\sim 1.0 \times 10^{-3}$. In addition, the conductance trace as a function of electrode spacing exhibits a convex downward curve having a minimum at $L_{es} = 26.46$ a.u. The present result of the conductance trace is consistent with the available experimental data

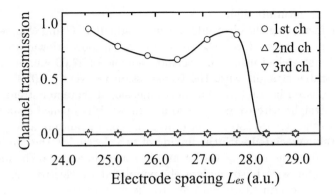

Fig. 7.4 Channel transmission at the Fermi level for the optimized three-aluminum-atom wire as a function of electrode spacing. Reprinted with permission from Ono and Hirose (2004b). © 2004 American Physical Society.

[Krans *et al.* (1993); Scheer *et al.* (1997); Cuevas *et al.* (1998); Mizobata *et al.* (2003a, 2003b)].

To gain insight into this unique conductance trace, we plot in Figs. 7.5 and 7.6 the electron-density distributions and current densities of the first channel at the Fermi level for several electrode spacings, respectively. The states incident from the left electrode are depicted. When the electrode spacing is $L_{es} = 24.57$ a.u. as shown in Figs. 7.5 (A) and 7.6 (A), the channel transmission is ~ 1. Incident electrons predominantly enter the bent wire along the atom rows in the [0$\bar{1}$1] and [101] directions, flow through the p_x and p_y orbitals of the center atom, and consequently form loop current (LC) rotating around the center atom. Electron transport through the wire in this stage is closely associated with the behavior of LC; in fact, when the center atom of the bent wire is forcibly displaced upward from the most stable position depicted in Fig. 7.5 (A) so as to draw up to the ($\bar{1}$11) plane occupied by the left edge atom of the wire, both LC and channel transmission increase due to the enhancement of current flow along the atom rows in the [0$\bar{1}$1] and [101] directions. Conversely, when the center atom is displaced downward, LC as well as channel transmission decreases. Moreover, as the center atom is further moved downward so that the wire becomes straight, electronic current becomes axially symmetric around the wire axis leading to the complete disappearance of LC, and the transmission becomes much smaller. In such a forcibly straightened wire, electrons heavily accumulate between the atoms since the interatomic distance is too small for a straightened wire to stabilize [Ono and Hirose (2003)], and incident electrons are reflected at the region where the electron density is unnaturally high.

Upon increasing the electrode spacing gradually, LC and channel transmission are suppressed because the center atom moves downward. When the wire is stretched up to an electrode spacing of 27.09 a.u., at which it forms a stable straight wire, the transmission recovers (~ 0.9) while LC disappears [see Fig. 7.6 (C)]. The one-dimensional character strengthens in a stable straight wire to give rise to an almost fully opened channel. According to the general aspect concerning the conductance of a very narrow constriction associated with the linking of two electrodes, the conductance is quantized in the units of G_0 for an ideal one-dimensional channel. The stable straight wire is so moderately populated by electrons as to have nearly perfect conductance, contrary to the forcibly straightened wire discussed above. Thus the first-channel transmission at the Fermi level is now larger than that of the bent wire of $L_{es} \geq 25.83$ a.u. After the wire distortion, there are no conduction channels between the two electrodes and only a small tunneling current of $\sim 10^{-3}$ G_0 is observed.

The present result that the conductance is ~ 1 G_0 is in accordance with that of the *ab initio* tight-binding calculation incorporating the square bases made of aluminum atoms [Palacios *et al.* (2002)]. One of the previous theoretical studies on the conductance of the three-aluminum-atom wire employing structureless jellium electrodes showed a similar current density [Kobayashi *et al.* (2000)]. It was implemented using an inferior pseudopotential made only of local parts and a very restricted model where the bond lengths of the wire are fixed to be 5.4 a.u. However, for a definite interpretation of the relationship between the geometry and conductance of the wire, a highly transferable pseudopotential having nonlocal parts and structural optimization are indispensable. Compared with other studies using nonlocal pseudopotentials and structureless jellium electrodes without any atomic layers or bases, most of the channel transmissions and conductance in the present study are smaller than those of the other studies: Lang (1995) and Kobayashi *et al.* (2001) reported that the conductance of the wire is ~ 1.5 G_0; moreover, Kobayashi *et al.* (2001) estimated that the transmissions of the second and third channels are ~ 0.3. In these calculations adopting jellium electrodes without any atomic layers, the channel transmission and conductance of the aluminum nanowire sensitively depend on direct effects from the jellium electrodes; therefore, the difference between our results and theirs may be mainly attributed to the absence of atomic layers or bases. For further details of this simulation, see Ono and Hirose (2004b).

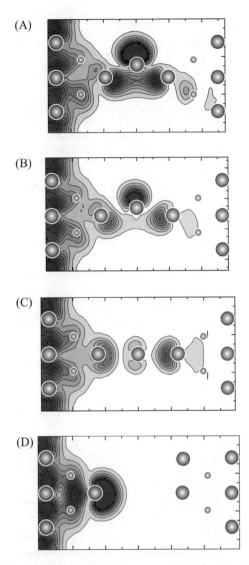

Fig. 7.5 Channel electron density distributions on the (110) plane at the Fermi level for the optimized three-aluminum-wire at (A) $L_{es} = 24.57$ a.u., (B) $L_{es} = 25.83$ a.u., (C) $L_{es} = 27.09$ a.u., and (D) $L_{es} = 28.35$ a.u. The states in the case of incident electrons coming from the left electrode are shown. The large and small spheres indicate the atomic positions on and below the cross section, respectively. Each contour represents 3.67×10^{-4} electron/bohr^3/eV higher(lower) than the adjacent contour lines, and the lowest contour is 3.67×10^{-4} electron/bohr^3/eV. Some of data taken from Ono and Hirose (2004b).

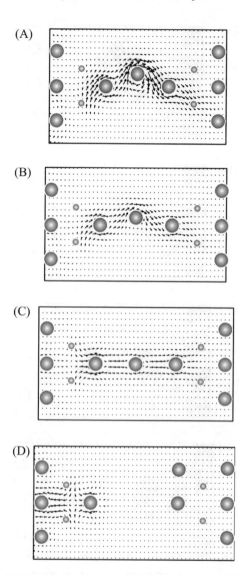

Fig. 7.6 Current density distributions on the (110) plane at the Fermi level for the optimized three-aluminum-wire at (A) L_{es} = 24.57 a.u., (B) L_{es} = 25.83 a.u., (C) L_{es} = 27.09 a.u., and (D) L_{es} = 28.35 a.u. The states in the case of incident electrons coming from the left electrode are shown. The large and small spheres indicate the atomic positions on and below the cross section, respectively. Some of data taken from Ono and Hirose. (2004b).

Chapter 8

Jellium Electrode Approximation

Tomoya Ono and Kikuji Hirose

8.1 Generalized Bloch States in Jellium Electrodes

Many first-principles studies have employed the *jellium* model in which
the ions are replaced by a uniform charge background of density in order to
save computational costs. We address the problem of obtaining the analytic
representation of generalized Bloch states in jellium electrodes within the
framework of the real-space finite-difference approach.

For the uniform background model, the Schrödinger equation is written
as

$$-\frac{1}{2}\nabla^2\phi + v\phi = E\phi, \tag{8.1}$$

where v is the constant potential arising from the uniform charge
background. The wave function ϕ can be factorized as $\phi(x, y, z) = \phi_x(x)\phi_y(y)\phi_z(z)$ in (8.1), where ϕ_x, ϕ_y, and ϕ_z are solutions of the one-
dimensional Schrödinger equation:

$$-\frac{1}{2}\frac{d^2}{dz^2}\phi_z(z) = \varepsilon_z\phi_z(z), \tag{8.2}$$

where $\varepsilon_z = E - v - \varepsilon_x - \varepsilon_y$. Here and hereafter, we consider the z direction
as an example. Discretizing the kinetic-energy operator in real space using
the central finite-difference formula of $N_f = 1$ [see Eq. (1.5)], we obtain

$$-\frac{1}{2}\frac{\phi_z(jh_z - h_z) - 2\phi_z(jh_z) + \phi_z(jh_z + h_z)}{h_z^2} = \varepsilon_z\phi_z(jh_z). \tag{8.3}$$

For the solution of (8.3), it is easily found that $\phi_z(jh_z \pm h_z) = e^{\pm i(k_z + i\kappa_z)h_z}\phi_z(jh_z)$, where k_z and κ_z are arbitrary real numbers. Then dividing all the terms of (8.3) by $\phi_z(jh_z)$, we obtain

$$\varepsilon_z = 2h_z^{-2} \sin^2 \frac{(k_z + i\kappa_z)h_z}{2}, \tag{8.4}$$

and $\phi_z(z) = e^{i(k_z + i\kappa_z)z}$. Here, either k_z or κ_z is zero since ε_z is a real number. The wave functions are right-propagating Bloch waves, left-propagating Bloch waves, rightward decreasing evanescent waves, and leftward decreasing evanescent waves in the cases of $k_z > 0$, $k_z < 0$, $\kappa_z > 0$, and $\kappa_z < 0$, respectively. When we impose a periodic boundary on wave functions, evanescent waves are not solutions of the Schrödinger equation. When we assume x and y to be the periodic directions parallel to the electrode surface, evanescent waves exponentially increase or decrease only in the z direction. Then, the generalized Bloch state deep inside semi-infinite jellium electrodes is written as

$$\phi(x, y, z) = e^{ik_x x} e^{ik_y y} e^{i(k_z + i\kappa_z)z}, \tag{8.5}$$

and the components of the columnar vectors $\Phi^A(z_k)$ $(A = in, ref$ and $tra)$ in (6.14) and (6.15) are expressed in the form of

$$\phi^A(x_i, y_j, z_k) = e^{ik_x x_i} e^{ik_y y_j} e^{i(k_z + i\kappa_z)z_k}. \tag{8.6}$$

(For more detailed treatment of wave vectors in the periodic x and y directions, see the discussion in Section 9.2.)

The wave-function matching based on the overbridging boundary-matching (OBM) scheme does not require the block tridiagonarity of the Hamiltonian in the transition region. Thus, when *jellium electrodes* are employed, any finite-difference formula can be adopted to realize the OBM procedure, not depending on whether the potentials between electrons and ions in the transition region are described in a local form or nonlocal one. See the comment in the case of employing jellium electrodes [Lang and Di Ventra (2003); Fujimoto and Hirose (2004)].

8.2 Application: Helical Gold Nanowires

So far, a large number of experiments concerning electron conduction through nanowires of a few nanometers in length have been implemented using a scanning tunneling microscope and a mechanically controllable break junction [see, for example, Krans (1993, 1995, 1996), Rubio *et al.* (1996), Scheer *et al.* (1997), Cuevas *et al.* (1998), Ohnishi *et al.* (1998), Yanson *et al.* (1998), Kizuka *et al.* (2001), Mizobata *et al.* (2003a, 2003b), and Smit *et al.* (2003)]. Among the most exciting discoveries are the helical gold nanowires (HGNs) with ∼1 nm diameter and ∼5 nm length observed by

Kondo and Takayanagi (2000) using transmission electron microscopy. The HGNs consist of several atom rows twisting around the nanowire axis and intervene between Au(011) electrodes. Although only geometrical structures of HGNs have been revealed in the transmission electron microscopy images and by theoretical calculations using empirical parameters, there remains much to be learned about their electronic characteristics; in particular, the conduction properties of HGNs are attracting considerable interest from both fundamental and practical points of view. Recently, Tosatti *et al.* (2001) examined geometrical and electronic structures of infinitely periodic helical wires using first-principles calculations. However, electron conduction properties of finite-length nanowires suspended between electrodes cannot be expected on the basis of the result of infinitely periodic wires because the scattering of electrons, which plays a crucial role in electron conduction, does not occur in infinitely periodic systems. Thus, to obtain a definite interpretation of the conduction properties, simulations using models of nanowires connected to electrodes are indispensable.

In this section, electron conduction properties of HGNs suspended between semi-infinite gold electrodes are studied based on first-principles calculations within the framework of the density functional theory. Figure 8.1 illustrates the models of the single-row, 7-1, 11-4, 13-6, 14-7-1, and 15-8-1 nanowires suspended between gold electrodes. A unit cell of 40.1 a.u. in the x and y directions with a periodic boundary, where the x and y directions are perpendicular to the nanowire axis, is employed. The atomic geometries of the HGNs are determined using the parameters reported by Kondo and Takayanagi (2000), and the lengths of the nanowires are set at $\sim$ 49.1 a.u. The electrode spacing is fixed at 51.2 a.u. even when the nanowires intervening between electrodes are replaced. The local pseudopotential of a gold 6s-valence electron constructed using the algorithm developed by Bachelet *et al.* [Hamann *et al.* (1979); Bachelet *et al.* (1982)] is employed to describe the ion-core potential and the central finite-difference formula, i.e., $N_f = 1$ in (1.5), is adopted for the kinetic-energy operator. We treat the exchange correlation effects with the local density approximation and imitate the gold electrodes with the structureless jellium ($r_s = 3.01$ a.u.). The cutoff energy is set to be 12 Ry, which corresponds to a grid spacing of 0.91 a.u. In addition, we use a higher cutoff energy of 110 Ry in the vicinity of nuclei with the augmentation of double-grid points.

We list, in Table 8.1, the conductance and channel transmissions of the gold nanowires, which are evaluated using the OBM method. In the case of the single-row nanowire, the number of conduction channels is one and the conductance is 1 G_0. This result concerning the conductance is in good agreement with those of previous experiments and theoretical calcu-

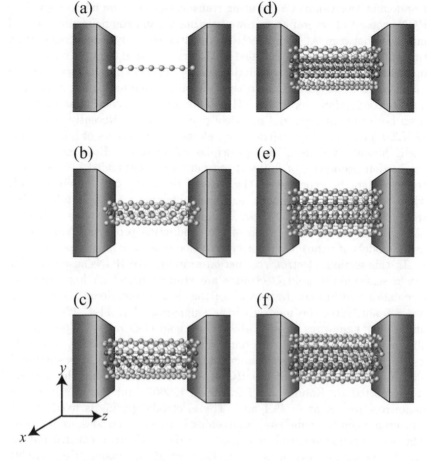

Fig. 8.1 Schematic descriptions of the scattering region of HGNs suspended between Au jellium electrodes: (a) single-row nanowire, (b) 7-1 nanowire, (c) 11-4 nanowire, (d) 13-6 nanowire, (e) 14-7-1 nanowire, and (f) 15-8-1 nanowire.

lations [Rubio *et al.* (1996); Ohnishi *et al.* (1998); Yanson *et al.* (1998); Okamoto and Takayanagi (1999); Fujimoto and Hirose (2003b); Smit *et al.* (2003)]. On the other hand, in the cases of the HGNs, the number of conduction channels does not correspond with the number of atom rows forming the HGNs. The number of channels is mainly affected by the geometrical and electronic structures of the nanowires rather than the interactions between electrodes and nanowires, since the number of channels in the 7-1 nanowire coincides with the number of energy bands crossing the

Table 8.1 Conductances and channel transmissions at the Fermi level.

	Single-row	7-1	11-4	13-6	14-7-1	15-8-1
Conductance (G_0)	0.96	5.19	9.08	11.97	13.82	14.44
1st ch.	0.958	0.997	1.000	0.996	1.000	1.000
2nd ch.	—	0.995	0.991	0.992	0.999	0.996
3rd ch.	—	0.970	0.986	0.982	0.991	0.991
4th ch.	—	0.938	0.890	0.978	0.984	0.982
5th ch.	—	0.653	0.884	0.962	0.959	0.975
6th ch.	—	0.640	0.874	0.958	0.947	0.954
7th ch.	—	—	0.753	0.940	0.940	0.942
8th ch.	—	—	0.748	0.878	0.927	0.931
9th ch.	—	—	0.620	0.872	0.893	0.899
10th ch.	—	—	0.517	0.744	0.888	0.891
11th ch.	—	—	0.497	0.735	0.881	0.841
12th ch.	—	—	0.318	0.706	0.867	0.834
13th ch.	—	—	—	0.618	0.770	0.732
14th ch.	—	—	—	0.383	0.700	0.691
15th ch.	—	—	—	0.200	0.542	0.645
16th ch.	—	—	—	0.002	0.527	0.595
17th ch.	—	—	—	—	—	0.537

Fermi level in the infinitely periodic helical gold wire. On the other hand, the channel transmissions are attributed to the interface between nanowires and electrodes; when we replaced the jellium electrodes with Au(011) crystalline electrodes, the conductance of the 7-1 nanowire becomes 4.8 G_0, whereas the number of conduction channel remains six. As a consequence, some channels are not fully open, i.e., their transmissions are less than 1. As a consequence, the conductance of the nanowire is much smaller than that expected from the results for the single-row nanowire, which is consistent with the recent experiment [Oshima *et al.* (2004)]. This result implies that the electron-conduction behaviors of the HGNs are different from those of the single-row nanowire. As an example, we show, in Fig. 8.2, the projection of the channel electron distributions onto the x–y plane in the 7-1 nanowires. The states incident from the left electrode are shown. One can easily recognize that conduction electrons do not flow along the atom rows in the 7-1 nanowire. The experimental consequences of these findings regarding the electron conduction behavior through the HGNs should be interesting to pursue. Thus, this situation will stimulate a new technology of electronic devices using atomic wires.

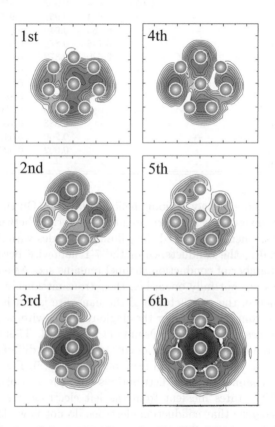

Fig. 8.2   Projection of the current charge density of the 7-1 wire onto the x–y plane.

8.3 Application: Fullerene Chains

Recently, special attention has been focused on atomic and molecular systems, such as organic molecules and fullerenes, because they are expected to be the ultimate size limit of functional devices and possess unique properties different from those of macroscopic systems. Joachim *et al.* [Joachim *et al.* (1995); Joachim and Gimzewski (1997)] indicated experimentally that a deformed C_{60} molecule has the possibility of being a nanoscale electrical amplifier. In addition, a C_{20} cage solely with pentagons has been produced from $C_{20}H_{20}$ [Prinzbach *et al.* (2000)], which is the smallest fullerene that is one of the candidates of much more minute electronic devices. While these prospects of fullerene-based devices are exciting, there remains much to be learned about their electronic characteristics.

On the theoretical side, first-principles calculations have been carried out to provide insight into geometric and electronic properties of nanostructures. Concerning C_{20} molecules, Miyamoto and Saito (2001) studied C_{20} one-dimensional infinite chains to indicate that some stable chain structures can be formed and one of them possesses semiconductivity. Later, electron-conduction properties of one-dimensional short chains made of several C_{20} molecules suspended between Al or Au leads were explored by Roland *et al.* (2001). They reported that a linear chain of such molecules acts primarily as metallic nanowires. Although these studies give us certain knowledge about fullerene-based devices, a transparent view on the quantized electron transport, which can be helpful to designing future devices, has not been provided yet, and further examinations of the conduction properties and the current flowing through the devices are indispensable.

In this section, we study electron transport properties, in particular, the conduction channels and their current distributions of C_{20} molecules suspended between two semi-infinite gold electrodes by first-principles calculations. The norm-conserving pseudopotentials [Kobayashi (1999a)] of Troullier and Martins (1991) and the local density approximation [Perdew and Zunger (1981)] of the density functional theory for exchange-correlation effects are used. Figure 8.3 shows the calculation models employed here.

We first explore the most stable geometry of C_{20} dimer among four types of dimers; a double-bonded, a single-bonded, CF1, and CF2 dimers. Then we concluded that the total energy of the double-bonded dimer is smaller than those of the single-bonded, CF1, and CF2 dimers, by 2.12, 0.88, and 2.60 eV, respectively. Here, the orientations of single-bonded and double-bonded dimers are in accordance with those proposed by Miyamoto and Saito (2001). In addition, CF1 and CF2 models are defined in Roland *et al.* (2001). We eventually adopt the two models, a double-bonded one shown in Fig. 8.3(a) and a single-bonded one shown in Fig. 8.3(b), for C_{20} monomers

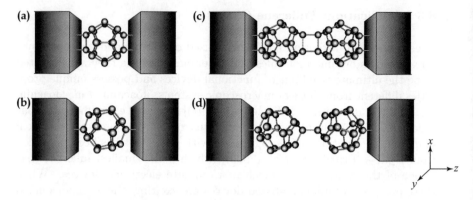

Fig. 8.3   Schematic descriptions of the scattering region of C_{20} molecules suspended between Au jellium electrodes: (a) double-bonded monomer, (b) single-bonded monomer, (c) double-bonded dimer, and (d) single-bonded dimer models.

sandwiched between electrodes. In the cases of dimers, we employ the double-bonded and single-bonded dimers indicated in Figs. 8.3(c) and (d), respectively. The C_{20} monomers and dimers are individually optimized, and then put between the electrodes. In order to reduce computational costs, we substitute jellium electrodes ($r_s = 3.01$ a.u.) for Au crystal ones. The distance between the edge atoms of inserted molecules and the jellium electrode is set at 0.91 a.u.

We then determine the stable structures of C_{20} monomers and dimers suspended between the Au electrodes. For structural optimization, a conventional supercell is employed under the periodic boundary condition in all directions. The size of the supercell is chosen to be $L_x = L_y = 21.6$ a.u. and $L_z = L_{mol} + 25$ a.u., where L_x and L_y are the lateral lengths of the supercell in the x and y directions parallel to the electrode surfaces, L_z is the length in the z direction, and L_{mol} is the length of the inserted molecules. A cutoff energy of 110 Ry, which corresponds to a grid spacing of 0.30 a.u., is taken, furthermore a higher cutoff energy of 987 Ry in the vicinity of nuclei is set with the augmentation of double-grid points. In the optimized geometries, the remaining forces acting on atoms are smaller than 1.65 nN. Consequently, all models become shorter along the z axis: the decreases of length are 2.9%, 13.5%, 2.1%, and 8.6% for the double-bonded monomer, single-bonded monomer, double-bonded dimer, and single-bonded dimer models, respectively.

Next, the electronic conductance and channel transmissions of the C_{20} molecules at the zero-bias limit are examined using the OBM method. We take a cutoff energy (a higher cutoff energy around nuclei) of 62 (555)

Table 8.2   Conductances and channel transmissions at the Fermi level: (a) dou-ble-bonded monomer, (b) single-bonded monomer, (c) double-bonded dimer, and (d) single-bonded dimer models. Data taken from Otani *et al.* (2004).

	Conductance(G_0)	Channel transimissions			
		1	2	3	4
(a)	1.57	0.47	0.47	0.35	0.20
(b)	0.83	0.36	0.18	0.17	0.06
(c)	0.18	0.16	0.01	0.00	0.00
(d)	0.17	0.07	0.06	0.03	0.01

Ry and employ the central finite-difference formula, i.e., $N_f = 1$ in (1.5). Table 8.2 collects the results of the conductances and channel transmissions at the Fermi level. Yet there is small band gap between highest occupied molecular orbital and lowest unoccupied molecular orbital in an isolated C_{20} molecule [Ivanov *et al.* (2001); Gianturco *et al.* (2002)], considerable electron conductions are observed in the C_{20} molecules suspended between the electrodes. The high conduction property is associated with a significant amount of charge transfer from the electrodes to molecules, which is localized at the interface between the electrodes and molecules. The increases of electrons per C_{20} molecule are 2.30%, 3.15%, 0.68%, and 0.24% for the double-bonded monomer, single-bonded monomer, double-bonded dimer, and single-bonded dimer models, respectively. As seen in Table 8.2, four channels that actually contribute to electronic conduction in both the monomer models: the conductances of the double-bonded one and the single-bonded one are 1.57 G_0 and 0.83 G_0, respectively. The difference in the conductance between the two monomer models is due to the number of atoms facing the jellium electrodes. In the case of the dimer models, the dimer models have the low conductance of 0.18 G_0 for the double-bonded one and 0.17 G_0 for the single-bonded one. Although more atoms facing the electrodes, the double-bonded dimer exhibits an approximately equal conductance to that of the single-bonded one, which cannot be expected from the above results of the monomer models. The result that conductance of the single-bonded dimer is higher than that of the double-bonded one is consistent with the previous theoretical study that the band gap of an infinite single-bond chain is significantly smaller than that of a double-bonded one [Miyamoto and Saito (2001)].

We show in Fig. 8.4 channel electron distributions at the Fermi level, where the cross sections at the center of the models along the z direction are depicted. Incident electrons are scattered at the entrances and exits of the molecules rather than within the C_{20} cages. What is more, electrons are clearly reflected at the C_{20}-C_{20} junctions in the case of dimers. In

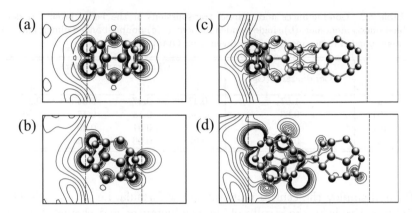

Fig. 8.4 Channel electron distributions at the Fermi level: (a) double-bonded monomer, (b) single-bonded monomer, (c) double-bonded dimer, and (d) single-bonded dimer models. The planes shown are perpendicular to the electrode surfaces and contain the atoms facing electrodes. The states incident from the left electrodes are depicted. The circles, lines, and broken lines represent carbon atoms, C-C bonds, and the edges of the jellium electrodes, respectively. Each contour represents 2.52×10^{-4} electron/bohr^3/eV higher(lower) than the adjacent contour lines, and the lowest contour is 2.52×10^{-4} electron/bohr^3/eV. Reprinted with permission from Otani *et al.* (2004). © 2004 American Physical Society.

Fig. 8.5, channel current distributions at the Fermi level are depicted, in which electrons are observed to conduct along the C-C bonds of the C_{20} cages rather than inside the cages. We found a more striking feature that local loop currents are induced around the outermost carbon atoms. For further details of this simulation, see Otani *et al.* (2004).

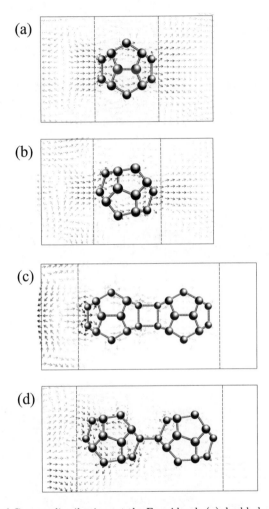

Fig. 8.5 Channel Current distributions at the Fermi level: (a) double-bonded monomer, (b) single-bonded monomer, (c) double-bonded dimer, and (d) single-bonded dimer models. The planes shown are the same as in Fig. 8.4. The circles, lines, and broken lines represent carbon atoms, C-C bonds, and the edges of the jellium electrodes, respectively. Reprinted with permission from Otani *et al.* (2004). © 2004 American Physical Society.

Chapter 9

Green's Function Formalism and the Overbridging Boundary-Matching Scheme

Yoshitaka Fujimoto, Tomoya Ono and Kikuji Hirose

9.1 Preliminary Remarks

We briefly discuss some basic properties of the Green's function for later convenience. The Green's function, which corresponds to a linear, Hermitian, time-independent differential operator, e.g., the Hamiltonian of a system $H(r, \nabla)$, and the complex variable $Z = E + i\eta$, is defined as the solution of the equation in the form of

$$[Z - H(r, \nabla)]G(r, r'; Z) = \delta(r - r') \tag{9.1}$$

subject to certain boundary conditions for r or r'. In the real-space finite-difference approach, the Hamiltonian and Green's function are described by the matrices $\hat{H}$ and $\hat{G}$, whose matrix elements are $H(r_l, r_{l'})$ and $G(r_l, r_{l'}; Z)$, respectively, and (9.1) is thus represented by

$$[Z - \hat{H}]\hat{G} = I \tag{9.2}$$

so that $\hat{G}$ is just the inverse matrix of $[Z - \hat{H}]$.

In a *finite* system, the Hamiltonian $H(r, \nabla)$ has discrete eigenvalues and therefore the Green's function exhibits only simple poles at the positions of the eigenvalues in the complex Z-plane:

$$G(r, r'; Z) = \sum_n \frac{\phi_n(r)\phi_n^*(r')}{Z - E_n}, \tag{9.3}$$

where $\{E_n\}$ is the set of eigenvalues and $\{\phi_n(r)\}$ is the complete orthonormal set of the eigenfunctions of $H(r, \nabla)$. We have already seen some examples of this type of the Green's function, e.g., Eqs. (6.11) and (7.8) for the truncated transition regions and (6.31) for the truncated bulk region.

We now consider the Green's functions of *infinitely* or *semi-infinitely* extending systems, and discuss common analytic properties of these Green's functions. The Hamiltonian $H(\boldsymbol{r}, \nabla)$ of such an extending system has a continuous spectrum $\{E_\lambda\}$, for instance, the bulk continuum in a periodic bulk, as well as a discrete spectrum $\{E_n\}$ (if any), so that the Green's function is written as

$$G(\boldsymbol{r}, \boldsymbol{r}'; Z) = \sum_n \frac{\phi_n(\boldsymbol{r})\phi_n^*(\boldsymbol{r}')}{Z - E_n} + \int \frac{\phi_\lambda(\boldsymbol{r})\phi_\lambda^*(\boldsymbol{r}')}{Z - E_\lambda} d\lambda. \qquad (9.4)$$

One can easily show that the expression (9.4) for G satisfies (9.1), using the completeness of the eigenfunction set $\{\phi_n, \phi_\lambda\}$,

$$\sum_n \phi_n(\boldsymbol{r})\phi_n^*(\boldsymbol{r}') + \int \phi_\lambda(\boldsymbol{r})\phi_\lambda^*(\boldsymbol{r}') d\lambda = \delta(\boldsymbol{r} - \boldsymbol{r}'). \qquad (9.5)$$

Since $H(\boldsymbol{r}, \nabla)$ is a Hermitian operator, all of its eigenvalues $\{E_n, E_\lambda\}$ are real. Hence, if Im $Z \neq 0$, then $Z \neq \{E_n, E_\lambda\}$, which means that $G(\boldsymbol{r}, \boldsymbol{r}'; Z)$ is an analytic function in the complex Z-plane, except at those points or portions $\{E_n, E_\lambda\}$ of the real Z-axis. It is noted that when Z is equal to E_λ which belongs to the continuous spectrum of $H(\boldsymbol{r}, \nabla)$, $G(\boldsymbol{r}, \boldsymbol{r}'; E_\lambda)$ is not well defined since the integrand in (9.4) has poles. Thus the continuous spectrum of $H(\boldsymbol{r}, \nabla)$ appears as a singular line along the real Z-axis. Nevertheless, one can attempt to define $G(\boldsymbol{r}, \boldsymbol{r}'; E_\lambda)$ by a limiting procedure. In the case of the ordinary Hamiltonian $H(\boldsymbol{r}, \nabla)$ whose eigenfunctions associated with the continuous spectrum are propagating wave functions, the side limits of $G(\boldsymbol{r}, \boldsymbol{r}'; E \pm i\eta)$ exist as a positive number η goes to zero ($\eta \to 0^+$) but are different from each other. Hence, the continuous spectrum of $H(\boldsymbol{r}, \nabla)$ produces a branch cut in $G(\boldsymbol{r}, \boldsymbol{r}'; Z)$ along the real Z-axis. We now define two Green's functions, which are referred to as the retarded Green's function (G^r) and the advanced Green's function (G^a),

$$G^r(\boldsymbol{r}, \boldsymbol{r}'; E) = \lim_{\eta \to 0^+} G(\boldsymbol{r}, \boldsymbol{r}'; E + i\eta) \qquad (9.6)$$

$$G^a(\boldsymbol{r}, \boldsymbol{r}'; E) = \lim_{\eta \to 0^+} G(\boldsymbol{r}, \boldsymbol{r}'; E - i\eta), \qquad (9.7)$$

and we give similar definitions for the corresponding matrices:

$$\hat{G}^r(E) = \lim_{\eta \to 0^+} (E + i\eta - \hat{H})^{-1} \qquad (9.8)$$

$$\hat{G}^a(E) = \lim_{\eta \to 0^+} G(E - i\eta - \hat{H})^{-1}. \tag{9.9}$$

From (9.4) one can easily see

$$G^a(\boldsymbol{r}, \boldsymbol{r}'; E) = [G^r(\boldsymbol{r}', \boldsymbol{r}; E)]^* \tag{9.10}$$

and equivalently,

$$\hat{G}^a(E) = \hat{G}^r(E)^\dagger. \tag{9.11}$$

Using the identity

$$\lim_{\eta \to 0^+} \frac{1}{E - E_\lambda + i\eta} = P\frac{1}{E - E_\lambda} - i\pi\delta(E - E_\lambda), \tag{9.12}$$

we obtain the following from (9.4) for the imaginary part of the diagonal elements of the retarded Green's function:

$$\mathrm{Im}\ G^r(\boldsymbol{r}, \boldsymbol{r}; E) = -\pi\sum_n |\phi_n(\boldsymbol{r})|^2\,\delta(E - E_n) - \pi\int |\phi_\lambda(\boldsymbol{r})|^2\,\delta(E - E_\lambda)d\lambda. \tag{9.13}$$

The right-hand side of (9.13) is proportional to

$$\rho(\boldsymbol{r}; E) = \sum_n |\phi_n(\boldsymbol{r})|^2\,\delta(E - E_n) + \int |\phi_\lambda(\boldsymbol{r})|^2\,\delta(E - E_\lambda)d\lambda \tag{9.14}$$

which is the local density of states (LDOS) per unit volume at energy E. Thus the LDOS can be expressed in terms of the diagonal elements of G^r:

$$\rho(\boldsymbol{r}; E) = -\frac{1}{\pi}\mathrm{Im}\ G^r(\boldsymbol{r}, \boldsymbol{r}; E). \tag{9.15}$$

A well-known example of G^r is the case of a free electron in the one-dimensional space; the basic equation (9.1) becomes

$$\left(Z + \frac{1}{2}\frac{d^2}{dz^2}\right)G(z, z'; Z) = \delta(z - z'), \tag{9.16}$$

from which we obtain the following expression of the retarded Green's function [Datta (1995)]. For $E > 0$,

$$G^r(z, z'; E) = \frac{1}{ik}e^{ik|z-z'|} \quad (k = \sqrt{2E} > 0). \tag{9.17}$$

For $E < 0$,

$$G^r(z, z'; E) = -\frac{1}{\kappa}e^{-\kappa|z-z'|} \quad (\kappa = \sqrt{-2E} > 0). \tag{9.18}$$

We note that the solutions (9.17) and (9.18), when regarding them as functions of z, represent an outgoing propagating wave $(E > 0)$ and a decreasing evanescent wave $(E < 0)$ which originate at the excitation point z', respectively, and further remark that (9.18) is obtainable by the substitution of $k = i\kappa$ in (9.17). Another example of G^r is illustrated in the next section.

9.2 Green's Functions in the Discretized Space within the Real-Space Finite-Difference Framework

In this section we give an explicit expression of the Green's function for a free electron in the discretized space. In addition, we discuss a peculiar property of the Green's functions at Z off the real energy axis for more general systems in the discretized space.

We begin by considering a one-dimensional free electron system within an interval on which the periodic boundary condition is imposed (box normalization). The space is divided into N_z equal-spacing grid points $\{z_l = lh_z\}$, $l = -N_z/2, ..., -1, 0, 1, ..., N_z/2 - 1$. Here h_z is the grid spacing, and for convenience, N_z is chosen to be an even number; when N_z is odd, l and n defined below are replaced by integers within the closed interval $[-(N_z - 1)/2, (N_z - 1)/2]$. Adopting the central finite-difference formula for the second-order derivative $\frac{d^2}{dz^2}$, we write the one-dimensional Schrödinger equation for a free electron

$$-\frac{1}{2}\frac{d^2}{dz^2}\phi_n(z) = \epsilon_n\phi_n(z) \tag{9.19}$$

as

$$-\frac{1}{2}\frac{\phi_n(z_{l-1}) - 2\phi_n(z_l) + \phi_n(z_{l+1})}{h_z^2} = \epsilon_n\phi_n(z_l). \tag{9.20}$$

It is easy to make sure that the solution of (9.20) under the periodic boundary condition is

$$\phi_n(z_l) = \frac{1}{\sqrt{N_z}}e^{iG_n z_l}, \tag{9.21}$$

which is the normalized eigenfunction belonging to the eigenvalue

$$\epsilon_n = h_z^{-2}(1 - \cos G_n h_z), \tag{9.22}$$

where G_n represents the inverse lattice vector $\{G_n = 2\pi n/h_z N_z\}$, $n = -N_z/2, ..., -1, 0, 1, ..., N_z/2 - 1$. Thus the Green's function in the present case is expressed as

$$G(z_l, z_{l'}; Z) = \frac{1}{N_z} \sum_{n=-N_z/2}^{N_z/2-1} \frac{e^{iG_n(z_l-z_{l'})}}{Z - \epsilon_n}$$

$$= \frac{h_z^2}{N_z} \sum_{n=-N_z/2}^{N_z/2-1} \frac{e^{i\frac{2\pi}{N_z}n(l-l')}}{h_z^2 Z - 1 + \cos\frac{2\pi}{N_z}n}. \qquad (9.23)$$

We next derive the Green's function for a free electron in the infinite discretized space by carrying out a limiting procedure $N_z \to \infty$ while keeping h_z constant in (9.23):

$$G(z_l, z_{l'}; Z) = \frac{h_z^2}{2\pi} \int_{-\pi}^{\pi} \frac{e^{i\theta(l-l')}}{h_z^2 Z - 1 + \cos\theta} d\theta. \qquad (9.24)$$

From (9.24) it is easily seen that this system has the continuous spectrum lying on the real E-axis such that $0 < E < 2h_z^{-2}$. To evaluate the integral, we observe that it depends on the absolute value $|l - l'|$, and we transform it to an integral over the complex variable ω $(= e^{i\theta})$ along the unit circle:

$$G(z_l, z_{l'}; Z) = \frac{h_z^2}{i\pi} \oint_{|\omega|=1} \frac{\omega^{|l-l'|}}{\omega^2 + 2\chi(Z)\omega + 1} d\omega$$

$$= \frac{h_z^2}{i\pi} \oint_{|\omega|=1} \frac{\omega^{|l-l'|}}{(\omega - \omega_1(Z))(\omega - \omega_2(Z))} d\omega, \qquad (9.25)$$

where

$$\chi(Z) = h_z^2 Z - 1$$
$$\omega_1(Z) = -\chi(Z) + \sqrt{\chi(Z)^2 - 1}$$
$$\omega_2(Z) = -\chi(Z) - \sqrt{\chi(Z)^2 - 1}. \qquad (9.26)$$

Here we obey the convention for the definition of the square root in which the imaginary part of $\sqrt{\omega}$ has the same sign as that of Im ω. It follows that $\omega_1\omega_2 = 1$. Only when the pole $\omega = \omega_1$ or ω_2 in (9.25) exists inside the unit circle in the ω-plane, does it contribute to the integral.

We here examine in detail the cases where Z goes to E on the real axis (i) inside and (ii) outside the continuous spectrum, by implementing a limiting procedure.

(i) Inside the continuous spectrum $(0 < E < 2h_z^{-2})$: To evaluate the retarded Green's function, we set $Z = E + i\eta$ $(\eta > 0)$, and easily find in (9.25) that only one pole $\omega = \omega_1$ (ω_2) enters into the inside of the unit circle in the case of $h_z^{-2} < E < 2h_z^{-2}$ $(0 < E < h_z^{-2})$. A usual evaluation of the integral by the method of residues is carried out in each case to yield the identical result for the retarded Green's function:

$$G^r(z_l, z_{l'}; E) = \frac{h_z^2 \left[1 - h_z^2 E + i\sqrt{h_z^2 E(2 - h_z^2 E)}\right]^{|l-l'|}}{i\sqrt{h_z^2 E(2 - h_z^2 E)}}$$

$$= \frac{h_z^2}{i\sin kh_z} e^{ik|z_l - z_{l'}|}, \tag{9.27}$$

where k $(0 < k < \pi h_z^{-1})$ is the wave number defined by

$$k = h_z^{-1} \cos^{-1}(1 - h_z^2 E). \tag{9.28}$$

(ii) Outside the continuous spectrum ($E < 0$ or $2h_z^{-2} < E$): By a similar estimation of the position of the pole in (9.25), we find that for $E < 0$ ($E > 2h_z^{-2}$) the pole $\omega = \omega_1$ (ω_2) contributes to the integral. Consequently, we have, for $E < 0$,

$$G^r(z_l, z_{l'}; E) = -\frac{h_z^2 \left[1 - h_z^2 E - \sqrt{h_z^2 E(h_z^2 E - 2)}\right]^{|l-l'|}}{\sqrt{h_z^2 E(h_z^2 E - 2)}}$$

$$= -\frac{h_z^2}{\sinh \kappa h_z} e^{-\kappa|z_l - z_{l'}|}, \tag{9.29}$$

and for $E > 2h_z^{-2}$,

$$G^r(z_l, z_{l'}; E) = \frac{h_z^2 \left[1 - h_z^2 E + \sqrt{h_z^2 E(h_z^2 E - 2)}\right]^{|l-l'|}}{\sqrt{h_z^2 E(h_z^2 E - 2)}}$$

$$= (-1)^{|l-l'|} \frac{h_z^2}{\sinh \kappa h_z} e^{-\kappa|z_l - z_{l'}|}, \tag{9.30}$$

where κ (> 0) is a parameter concerning the exponent of an evanescent wave, given by

$$\kappa = h_z^{-1} \cosh^{-1} \left|1 - h_z^2 E\right|. \tag{9.31}$$

Using (9.27) – (9.31) with (9.15), the LDOS per grid point is evaluated in the form of

$$\rho(z_l; E) = \begin{cases} \dfrac{h_z^2}{\pi\sqrt{h_z^2 E(2 - h_z^2 E)}} & \text{for } 0 < E < 2h_z^{-2} \\[2ex] 0 & \text{for } \textit{otherwise.} \end{cases} \tag{9.32}$$

We note that the retarded Green's function and LDOS, (9.27) – (9.32), correspond to those of a one-dimensional simple lattice in the tight-binding model with nearest-neighbor interactions.

It is straightforward to extend the above argument to the three-dimensional case. We deal with the model of a free electron in which the discretized space is infinite in the z direction and periodic in the x and y directions. The grid points are denoted by $r_l = (x_{l_x}, y_{l_y}, z_{l_z}) = (l_x h_x, l_y h_y, l_z h_z)$, where $l_\mu = -N_\mu/2, ..., -1, 0, 1, ..., N_\mu/2 - 1$, and h_μ and N_μ are the grid spacing and the total number of the grid points in the μ direction ($\mu = x, y, z$), respectively, but $N_z \to \infty$ so that l_z can be any integer. We have the following expression for the Green's function, similar to (9.25) in the one-dimensional case:

$$G(r_l, r_{l'}; Z, k_{||}) = \frac{h_z^2}{i\pi N_x N_y} \sum_{n_x=-N_x/2}^{N_x/2-1} \sum_{n_y=-N_y/2}^{N_y/2-1} e^{i(k_{||}+G_{||n})(r_{||l}-r_{||l'})}$$

$$\times \oint_{|\omega|=1} \frac{\omega^{|l_z-l'_z|}}{\omega^2 + 2\tilde{\chi}(Z, k_{||}, G_{||n})\omega + 1} d\omega, \qquad (9.33)$$

where $r_{||l} = (x_{l_x}, y_{l_y})$ is the lateral coordinates, $k_{||} = (k_x, k_y)$ is the lateral Bloch wave vector within the first Brillouin zone, $G_{||n}$ is the lateral inverse lattice vector, i.e.,

$$G_{||n} = (G_{n_x}, G_{n_y}) = \left(\frac{2\pi}{h_x N_x} n_x, \frac{2\pi}{h_y N_y} n_y \right), \qquad (9.34)$$

and

$$\tilde{\chi}(Z, k_{||}, G_{||n}) = h_z^2 Z - 3 + \cos(k_x + G_{n_x})h_x + \cos(k_y + G_{n_y})h_y. \qquad (9.35)$$

The integral in (9.33) can be analytically calculated in an analogous manner to the derivation of (9.27) – (9.31) from (9.25). Consequently, the retarded Green's function in the energy range of $0 < E < 2h_z^{-2}$ is written as

$$G^r(r_l, r_{l'}; E, k_{||}) = \frac{h_z^2}{i N_x N_y} \sum_{n_x=-N_x/2}^{N_x/2-1} \sum_{n_y=-N_y/2}^{N_y/2-1} e^{i(k_{||}+G_{||n})(r_{||l}-r_{||l'})}$$

$$\times \frac{1}{\sin k_z h_z} e^{ik_z \left| z_{l_z} - z_{l'_z} \right|}. \qquad (9.36)$$

Here,

$$k_z = h_z^{-1} \cos^{-1} \left[-\tilde{\chi}(E, k_{||}, G_{||n}) \right], \qquad (9.37)$$

which depends on E, $k_{||}$ and $G_{||n}$. If $\left| \tilde{\chi}(E, k_{||}, G_{||n}) \right| < 1$, then k_z ranges from 0 to πh_z^{-1} and corresponds to the z-component of the wave vector of an outwardly propagating wave; otherwise, k_z is analytically continued to κ_z (> 0) such that $k_z = i\kappa_z$, where

$$\kappa_z = h_z^{-1} \cosh^{-1} \left[-\tilde{\chi}(E, \boldsymbol{k}_{||}, \boldsymbol{G}_{||n}) \right] \qquad (9.38)$$

accounts for the exponential factor of an evanescent wave which decreases as $\left| z_{l_z} - z_{l'_z} \right|$ becomes large. This analytic representation of the retarded Green's function for a free electron can be utilized in the analysis based on the Lippmann–Schwinger equation discussed in the next chapter.

We move on to the discussion of Green's functions at $Z \neq real$ for general crystalline bulks. It is worth observing that in (9.25) the off-diagonal matrix elements $G(z_l, z_{l'}; Z \neq \text{real})$ for a free electron system must decay exponentially with distance $|l - l'|$ because $\omega_i(Z)$ ($i = 1$ or 2), which corresponds to the position of the pole contributing to the integral in (9.25), is such that $|\omega_i(Z)| \lessgtr 1$ for Z off the real energy axis. In the following theorem, we show that this decaying property is a common characteristic of off-real-energy Green's functions of more general cases of systems [Lee and Joannopoulos (1981b)]. It will be seen in the next section that this argument, combined with Theorem 6.5 (see Section 6.5), can lead to Theorem 9.2 that gives an explicit expression of the retarded Green's functions, one of the principal ingredients of the Green's function formalism.

Theorem 9.1 *For the three-dimensional crystalline bulk depicted in Fig. 6.4,*

$$G(z_l^{M+p}, z_{l'}^M; Z \neq real, \boldsymbol{k}_{||}) \to 0 \; as \; |p| \to \infty, \qquad (9.39)$$

where p is an integer. [See Fig. 6.4 for the geometry of the system. In denoting block-matrix elements of the Green's function in (9.39), we follow the notation used in Chapter 6, e.g., see $\mathcal{G}_T^M$ in (6.32).]

Proof. We give the proof of (9.39), according to Lee and Joannopoulos. In a usual supercell treatment of a crystalline bulk, it is known that the block-matrix element of G within the M-th unit cell is expressed as

$$G(z_l^M, z_{l'}^M; Z, \boldsymbol{k}_{||}) = \frac{L_z}{2\pi} \int_{1BZ} dk_z \sum_n \frac{\Phi_n(z_l^M; \boldsymbol{k}_{||}, k_z)\Phi_n^\dagger(z_{l'}^M; \boldsymbol{k}_{||}, k_z)}{Z - E_n(\boldsymbol{k}_{||}, k_z)},$$

$$(9.40)$$

where E_n and Φ_n are the eigenvalues and eigenvectors of the periodic Hamiltonian $\hat{H}_{per}^M(\omega = e^{ik_z L_z}, \boldsymbol{k}_{||})$ of (6.39), respectively; the symbol '$1BZ$' means that the integration must be restricted to the first Brillouin zone of k_z, i.e., $-\pi/L_z \leq k_z < \pi/L_z$. According to the convention used in Chapter 6, $\Phi_n(z_l^M; \boldsymbol{k}_{||}, k_z)$ stands for an N-dimensional columnar vector constructed by $\{\phi_n(x_i, y_j, z_l^M; \boldsymbol{k}_{||}, k_z)\}$ on the x–y plane at $z = z_l^M$. Since the integrand in (9.40) is the block-matrix element of $[Z - \hat{H}_{per}^M]^{-1}$ associated with the planes at $z = z_l^M$ and $z_{l'}^M$, we write it explicitly as

$$\sum_n \frac{\Phi_n(z_l^M; \boldsymbol{k}_{||}, k_z)\Phi_n^\dagger(z_{l'}^M; \boldsymbol{k}_{||}, k_z)}{Z - E_n(\boldsymbol{k}_{||}, k_z)} = \left(z_l^M \left\| \left[Z - \hat{H}_{per}^M(e^{ik_z L_z}, \boldsymbol{k}_{||}) \right]^{-1} \right\| z_{l'}^M \right).$$

(9.41)

Taking into account the Bloch condition

$$\Phi_n(z_l^{M+p}; \boldsymbol{k}_{||}, k_z) = e^{ipk_z L_z} \Phi_n(z_l^M; \boldsymbol{k}_{||}, k_z),$$

(9.42)

one sees that the block-matrix element of the Green's function G associated with the planes at $z = z_l^{M+p}$ and $z_{l'}^M$ is given by

$$G(z_l^{M+p}, z_{l'}^M; Z, \boldsymbol{k}_{||})$$
$$= \frac{L_z}{2\pi} \int_{1BZ} dk_z \, e^{ipk_z L_z} \left(z_l^M \left\| \left[Z - \hat{H}_{per}^M(e^{ik_z L_z}, \boldsymbol{k}_{||}) \right]^{-1} \right\| z_{l'}^M \right). \quad (9.43)$$

The change of the variable from k_z to $\omega = e^{ik_z L_z}$ transforms (9.43) into

$$G(z_l^{M+p}, z_{l'}^M; Z, \boldsymbol{k}_{||}) = \frac{1}{2\pi i} \oint_{|\omega|=1} d\omega \, \omega^{p-1} \left(z_l^M \left\| \left[Z - \hat{H}_{per}^M(\omega, \boldsymbol{k}_{||}) \right]^{-1} \right\| z_{l'}^M \right).$$

(9.44)

We first examine the case of a positive integer p. Making use of the well-known formula that expresses the inverse of matrix X in the form of

$$X^{-1} = Cof\,[X]/|X|,$$

(9.45)

in which $Cof\,[X]$ is a matrix constructed by the cofactors associated with X, and subsequently, employing (6.87), (6.91), (6.92) and (6.96), one obtains

$$\left[Z - \hat{H}_{per}^M(\omega, \boldsymbol{k}_{||}) \right]^{-1}$$

$$= \frac{Cof\left[Z - \hat{H}_{per}^M(\omega, \boldsymbol{k}_{||}) \right]}{\left| Z - \hat{H}_{per}^M(\omega, \boldsymbol{k}_{||}) \right|}$$

$$= \frac{(-\omega)^N Cof\left[Z - \hat{H}_{per}^M(\omega, \boldsymbol{k}_{||}) \right]}{\left| Z - \hat{H}_T^M(\boldsymbol{k}_{||}) \right| \left| \Pi_2(Z, \boldsymbol{k}_{||}) \right| \left| \Pi_1(Z, \boldsymbol{k}_{||})\Pi_2(Z, \boldsymbol{k}_{||})^{-1} - \omega \right|}$$

$$= \frac{(-\omega)^N Cof\left[Z - \hat{H}_{per}^M(\omega, \boldsymbol{k}_{||}) \right]}{\prod_{l=1}^m |B(z_l^M)| \cdot \prod_{n=1}^{2N} (\omega - \lambda_n(Z, \boldsymbol{k}_{||}))}. \quad (9.46)$$

Here, $\{\lambda_n(Z, \boldsymbol{k}_{||})\}$, $n = 1, 2, ..., 2N$, are eigenvalues of the generalized eigenvalue equation (6.85). Since $(-\omega)^N (z_l^M | Cof[Z - \hat{H}_{per}^M(\omega, \boldsymbol{k}_{||})] | z_{l'}^M)$ contains only positive powers of ω [see Eq. (6.39)], we can expand it to be

$$(-\omega)^N \left(z_l^M \left| Cof \left[Z - \hat{H}_{per}^M(\omega, \boldsymbol{k}_{||}) \right] \right| z_{l'}^M \right) = \sum_{k=0}^{2N} \omega^k D_k(z_l^M, z_{l'}^M; Z, \boldsymbol{k}_{||}),$$

(9.47)

where

$$
\begin{aligned}
&D_k(z_l^M, z_{l'}^M; Z, \boldsymbol{k}_{||}) \\
&= \frac{1}{k!} \frac{\partial^k}{\partial \omega^k} \left\{ (-\omega)^N \left(z_l^M \left| Cof \left[Z - \hat{H}_{per}^M(\omega, \boldsymbol{k}_{||}) \right] \right| z_{l'}^M \right) \right\} \Bigg|_{\omega=0}.
\end{aligned}
$$

(9.48)

Equation (9.44) is then expressed as

$$
\begin{aligned}
G(z_l^{M+p}, z_{l'}^M; Z, \boldsymbol{k}_{||}) &= \frac{1}{\prod_{l=1}^m |B(z_l^M)|} \sum_{k=0}^{2N} D_k(z_l^M, z_{l'}^M; Z, \boldsymbol{k}_{||}) \\
&\times \frac{1}{2\pi i} \oint_{|\omega|=1} d\omega \frac{\omega^{p+k-1}}{\prod_{n=1}^{2N} (\omega - \lambda_n(Z, \boldsymbol{k}_{||}))}.
\end{aligned}
$$

(9.49)

Only the pole $\omega = \lambda_n(Z, \boldsymbol{k}_{||})$ existing inside the unit circle in the ω-plane contributes to the integral. We already know that in Theorem 6.5 $|\lambda_n(Z, \boldsymbol{k}_{||})| \neq 1$ and there are equal numbers of $\lambda_n(Z, \boldsymbol{k}_{||})$ with modulus greater and less than 1 for Z off the real energy axis ($Z \neq E$). Therefore, for a positive integer p, we have

$$
\begin{aligned}
G(z_l^{M+p}, z_{l'}^M; Z, \boldsymbol{k}_{||}) &= \frac{1}{\prod_{l=1}^m |B(z_l^M)|} \sum_{k=0}^{2N} D_k(z_l^M, z_{l'}^M; Z, \boldsymbol{k}_{||}) \\
&\times \sum_{\substack{\nu=1 \\ |\lambda_\nu|<1}}^{2N} \frac{(\lambda_\nu(Z, \boldsymbol{k}_{||}))^{p+k-1}}{\prod_{n \neq \nu}^{2N} (\lambda_\nu(Z, \boldsymbol{k}_{||}) - \lambda_n(Z, \boldsymbol{k}_{||}))}.
\end{aligned}
$$

(9.50)

This equation indicates that $G(z_l^{M+p}, z_{l'}^M; Z, \boldsymbol{k}_{||})$ decays as $p \to \infty$.

We next turn to the case of a negative integer p. Let us change the variable from ω to $1/\omega$ in (9.44). This gives

$$
\begin{aligned}
&G(z_l^{M+p}, z_{l'}^M; Z, \boldsymbol{k}_{||}) \\
&= \frac{1}{2\pi i} \oint_{|\omega|=1} d\omega \, \omega^{|p|-1} \left(z_l^M \left| \left[Z - \hat{H}_{per}^M \left(\frac{1}{\omega}, \boldsymbol{k}_{||} \right) \right]^{-1} \right| z_{l'}^M \right).
\end{aligned}
$$

(9.51)

Similarly to (9.46), putting $\left[Z - \hat{H}_{per}^M (1/\omega, \boldsymbol{k}_{||})\right]^{-1}$ into the form of

$$
\begin{aligned}
\left[Z - \hat{H}_{per}^M (\tfrac{1}{\omega}, \boldsymbol{k}_{||})\right]^{-1} &\\
&= \frac{(-\omega)^N Cof \left[Z - \hat{H}_{per}^M \left(\tfrac{1}{\omega}, \boldsymbol{k}_{||}\right)\right]}{\left|Z - \hat{H}_T^M(\boldsymbol{k}_{||})\right| \left|\Pi_1(z, \boldsymbol{k}_{||})\right| \left|\Pi_2(Z, \boldsymbol{k}_{||})\Pi_1(Z, \boldsymbol{k}_{||})^{-1} - \omega\right|} \\
&= \frac{(-\omega)^N Cof \left[Z - \hat{H}_{per}^M \left(\tfrac{1}{\omega}, \boldsymbol{k}_{||}\right)\right]}{\prod_{l=1}^m \left|B(z_l^M)\right|^* \cdot \prod_{n=1}^{2N} \left(\omega - \lambda_n(z, \boldsymbol{k}_{||})^{-1}\right)}
\end{aligned}
\tag{9.52}
$$

by the use of (6.97), and making the expansion as

$$
(-\omega)^N \left(z_l^M \left|Cof \left[Z - \hat{H}_{per}^M \left(\tfrac{1}{\omega}, \boldsymbol{k}_{||}\right)\right]\right| z_{l'}^M \right) = \sum_{k=0}^{2N} \omega^k D_k'(z_l^M, z_k^M; Z, \boldsymbol{k}_{||}),
\tag{9.53}
$$

where $\{D_k'\}$ are expansion coefficients similar to $\{D_k\}$ of (9.48), one obtains the following expression for (9.51):

$$
\begin{aligned}
G(z_l^{M+p}, z_{l'}^M; Z, \boldsymbol{k}_{||}) &= \frac{1}{\prod_{l=1}^m \left|B(z_l^M)\right|^*} \sum_{k=0}^{2N} D_k'(z_l^M, z_{l'}^M; Z, \boldsymbol{k}_{||}) \\
&\quad \times \frac{1}{2\pi i} \oint_{|\omega|=1} d\omega \frac{\omega^{|p|+k-1}}{\prod_{n=1}^{2N} \left(\omega - \lambda_n(Z, \boldsymbol{k}_{||})^{-1}\right)}.
\end{aligned}
\tag{9.54}
$$

Thus, carrying out the integration, one finds

$$
\begin{aligned}
G(z_l^{M+p}, z_{l'}^M; Z, \boldsymbol{k}_{||}) &= \frac{1}{\prod_{l=1}^m \left|B(z_l^M)\right|^*} \sum_{k=0}^{2N} D_k'(z_l^M, z_{l'}^M; Z, \boldsymbol{k}_{||}) \\
&\quad \times \sum_{\substack{\nu=1 \\ |\lambda_\nu|>1}}^{2N} \frac{(\lambda_n(Z, \boldsymbol{k}_{||})^{-1})^{|p|+k-1}}{\prod_{n\neq\nu}^{2N} \left(\lambda_\nu(Z, \boldsymbol{k}_{||})^{-1} - \lambda_n(Z, \boldsymbol{k}_{||})^{-1}\right)},
\end{aligned}
\tag{9.55}
$$

which shows that $G(z_l^{M+p}, z_{l'}^M; Z, \boldsymbol{k}_{||})$ diminishes as $p \to -\infty$. $\qquad\square$

9.3 Green's Function of a Whole System Including the Transition Region and Two Semi-Infinite Electrodes

We deal with the Green's function of a system composed of the transition region connected to two semi-infinite crystalline electrodes, as shown in Fig. 9.1, within the framework of the real-space finite-difference approach. We assume two-dimensional periodicity in the x and y directions, and use a generalized z-coordinate ζ_l instead of z_l; thus, the formalism includes norm-conserving pseudopotential techniques (see Section 7.1).

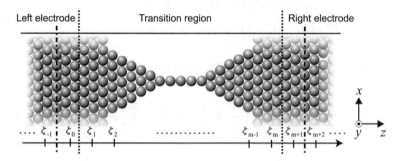

Fig. 9.1 Sketch of a system with the transition region intervening between the left and right semi-infinite crystalline electrodes. The dash-dotted lines correspond to the borders of the partitioning of the Hamiltonian matrix in (9.56) and Fig. 9.2, whereas the dotted lines denote the borders of the individual regions which were used in wave-function matching (see Fig. 7.2).

The left and right electrodes are considered to couple only indirectly via the transition region, and therefore, the Hamiltonian matrix of the whole system, $\hat{H}(\mathbf{k}_{||})$, can be partitioned into submatrices that correspond to the individual regions (see Fig. 9.2):

$$\hat{H}(\mathbf{k}_{||}) = \left[\begin{array}{c|c|c} \hat{H}_L(\mathbf{k}_{||}) & \hat{B}_{LT} & 0 \\ \hline \hat{B}_{LT}^{\dagger} & \hat{H}_T(\mathbf{k}_{||}) & \hat{B}_{TR} \\ \hline 0 & \hat{B}_{TR}^{\dagger} & \hat{H}_R(\mathbf{k}_{||}) \end{array} \right], \qquad (9.56)$$

where the borders of the partitioning of $\hat{H}$ are drawn according to the dash-dotted lines in Fig. 9.1; the submatrix $\hat{H}_T$ includes the matrix elements in the transition region, $\hat{H}_L$ ($\hat{H}_R$) relates to the semi-infinite left- (right-)electrode region, and $\hat{B}_{LT}$ ($\hat{B}_{TR}$) stands for 'a coupling matrix' which is nonzero only for some connection area between the transition region and the left (right) electrode. We assume the block tridiagonality of $\hat{H}_L$ ($\hat{H}_R$), all of the block-matrix elements of which are N dimensional,

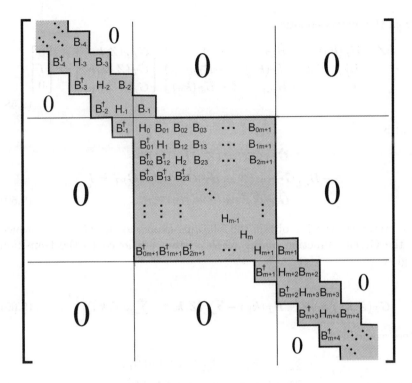

Fig. 9.2 Partitioning of the Hamiltonian matrix, $\hat{H}$ of (9.56), associated with the whole system sketched in Fig. 9.1. The partition lines are identical to those in (9.56).

and also assume such a simple structure of $\hat{B}_{LT}$ $(\hat{B}_{TR})$ that only one N-dimensional block-matrix element $B(\zeta_{-1})$ $(B(\zeta_{m+1}))$ is nonzero, as illustrated in Fig. 9.2, in which block-matrix elements H_l, B_l and $B_{ll'}$ are the abbreviations of $H(\zeta_l, \boldsymbol{k}_{||})$, $B(\zeta_l)$ and $B(\zeta_l, \zeta_{l'})$, respectively. On the other hand, $\hat{H}_T$ in the transition region is treated as a general nonsparse matrix here and in the next section.

By definition, the Green's function of the whole system is written as

$$\hat{G}(Z, \boldsymbol{k}_{||}) = \begin{bmatrix} \hat{G}_L(Z, \boldsymbol{k}_{||}) & \hat{G}_{LT}(Z, \boldsymbol{k}_{||}) & \hat{G}_{LR}(Z, \boldsymbol{k}_{||}) \\ \hat{G}_{TL}(Z, \boldsymbol{k}_{||}) & \hat{G}_T(Z, \boldsymbol{k}_{||}) & \hat{G}_{TR}(Z, \boldsymbol{k}_{||}) \\ \hat{G}_{RL}(Z, \boldsymbol{k}_{||}) & \hat{G}_{RT}(Z, \boldsymbol{k}_{||}) & \hat{G}_R(Z, \boldsymbol{k}_{||}) \end{bmatrix}$$

$$= \begin{bmatrix} Z - \hat{H}_L(\boldsymbol{k}_{||}) & -\hat{B}_{LT} & 0 \\ -\hat{B}_{LT}^\dagger & Z - \hat{H}_T(\boldsymbol{k}_{||}) & -\hat{B}_{TR} \\ 0 & -\hat{B}_{TR}^\dagger & Z - \hat{H}_R(\boldsymbol{k}_{||}) \end{bmatrix}^{-1}. \quad (9.57)$$

From the matrix equation

$$
\begin{bmatrix}
Z - \hat{H}_L(\mathbf{k}_{||}) & -\hat{B}_{LT} & 0 \\
-\hat{B}_{LT}^\dagger & Z - \hat{H}_T(\mathbf{k}_{||}) & -\hat{B}_{TR} \\
0 & -\hat{B}_{TR}^\dagger & Z - \hat{H}_R(\mathbf{k}_{||})
\end{bmatrix}
\begin{bmatrix}
\hat{G}_{LT}(Z, \mathbf{k}_{||}) \\
\hat{G}_T(Z, \mathbf{k}_{||}) \\
\hat{G}_{RT}(Z, \mathbf{k}_{||})
\end{bmatrix}
=
\begin{bmatrix}
0 \\
I \\
0
\end{bmatrix},
\tag{9.58}
$$

that is,

$$
\hat{G}_{LT}\hat{G}_T^{-1} = (Z - \hat{H}_L)^{-1}\hat{B}_{LT}
$$
$$
-\hat{B}_{LT}^\dagger\hat{G}_{LT} + (Z - \hat{H}_T)\hat{G}_T - \hat{B}_{TR}\hat{G}_{RT} = I
$$
$$
\hat{G}_{RT}\hat{G}_T^{-1} = (Z - \hat{H}_R)^{-1}\hat{B}_{TR}^\dagger,
\tag{9.59}
$$

it is straightforward to obtain the following expression for $\hat{G}_T(Z, \mathbf{k}_{||})$ which is the Green's function of the whole system portioned to the transition region:

$$
\hat{G}_T(Z, \mathbf{k}_{||}) = \left[Z - \hat{H}_T(\mathbf{k}_{||}) - \hat{\Sigma}_L(Z, \mathbf{k}_{||}) - \hat{\Sigma}_R(Z, \mathbf{k}_{||}) \right]^{-1}.
\tag{9.60}
$$

Here we defined

$$
\hat{\Sigma}_L(Z, \mathbf{k}_{||}) = \hat{B}_{LT}^\dagger\hat{\mathcal{G}}_L(Z, \mathbf{k}_{||})\hat{B}_{LT}
$$
$$
\hat{\Sigma}_R(Z, \mathbf{k}_{||}) = \hat{B}_{TR}\hat{\mathcal{G}}_R(Z, \mathbf{k}_{||})\hat{B}_{TR}^\dagger
\tag{9.61}
$$

and refer to these matrices as the self-energy terms on the basis of the nomenclature of the corresponding terms defined in the quantum field theory, where

$$
\hat{\mathcal{G}}_L(Z, \mathbf{k}_{||}) = \left[Z - \hat{H}_L(\mathbf{k}_{||}) \right]^{-1}
$$
$$
\hat{\mathcal{G}}_R(Z, \mathbf{k}_{||}) = \left[Z - \hat{H}_R(\mathbf{k}_{||}) \right]^{-1}
\tag{9.62}
$$

are the Green's functions of the semi-infinite left and right electrodes with right- and left-side truncations, respectively. [The script capital letter $\hat{\mathcal{G}}$ is used for describing the Green's function of a semi-infinite system with one-side truncation as well as an isolated (two-side truncated) system; this may prevent confusing $[Z - \hat{H}_A]^{-1}$ with $\hat{G}_A$ defined by (9.57), where $A = L, T$ and R.] From (9.60), one sees that the Green's function associated with the isolated transition-region Hamiltonian $\hat{H}_T(\mathbf{k}_{||})$, namely,

$$
\hat{\mathcal{G}}_T(Z, \mathbf{k}_{||}) = \left[Z - \hat{H}_T(\mathbf{k}_{||}) \right]^{-1},
\tag{9.63}
$$

is so extended to $\hat{G}_T(Z, \mathbf{k}_{||})$ as to include the effects of the semi-infinite electrodes through the self-energy terms $\sum_{\{L,R\}}(Z, \mathbf{k}_{||})$. Recall that $\hat{\mathcal{G}}_T$ is the Green's function that was introduced for wave-function matching in Section 7.1 [see Eq. (7.8)]. It should be noted that (9.60) is equivalent to Dyson's equation in the standard form [Williams *et al.* (1982)]:

$$\hat{G}_T(Z, \mathbf{k}_{||}) = \hat{\mathcal{G}}_T(Z, \mathbf{k}_{||}) + \hat{\mathcal{G}}_T(Z, \mathbf{k}_{||}) \left[\hat{\sum}_L(Z, \mathbf{k}_{||}) + \hat{\sum}_R(Z, \mathbf{k}_{||}) \right] \hat{G}_T(Z, \mathbf{k}_{||}),$$
(9.64)

in which the unperturbed Green's function $\hat{\mathcal{G}}_T(Z, \mathbf{k}_{||})$ is modified into the perturbed one $\hat{G}_T(Z, \mathbf{k}_{||})$ through the perturbation $\hat{\sum}_{\{L,R\}}(Z, \mathbf{k}_{||})$. In the next section, we give the exact analytic solution of Deyson's equation (9.64) [see Eqs. (9.93) – (9.96)].

The rest of this section is devoted to the evaluation of the retarded Green's functions of the left- and right-electrode regions,

$$\hat{\mathcal{G}}^r_{\{L,R\}}(E, \mathbf{k}_{||}) = \lim_{\eta \to 0^+} \hat{\mathcal{G}}_{\{L,R\}}(E + i\eta, \mathbf{k}_{||}),$$
(9.65)

and hence the retarded self-energies

$$\hat{\sum}{}^r_{\{L,R\}}(E, \mathbf{k}_{||}) = \lim_{\eta \to 0^+} \hat{\sum}_{\{L,R\}}(E + i\eta, \mathbf{k}_{||}).$$
(9.66)

The retarded Green's function $\hat{G}^r_T(E, \mathbf{k}_{||})$ derived from (9.60) [or (9.64)] will be discussed in the next section.

From (9.61), we find that since $\hat{B}_{LT}$ ($\hat{B}_{TR}$) is assumed to have only one nonzero block-matrix element $B(\zeta_{-1})$ ($B(\zeta_{m+1})$) [see Eq. (9.56) and Fig. 9.2], the retarded self-energy terms (9.61) take the very simple form of

$$\hat{\sum}{}^r_L(E, \mathbf{k}_{||}) = \begin{bmatrix} \sum_L^r(\zeta_0; E, \mathbf{k}_{||}) & 0 & \cdots & 0 \\ 0 & & 0 & \cdots & 0 \\ \vdots & & & \vdots \\ 0 & & 0 & \cdots & 0 \end{bmatrix}$$

$$\hat{\sum}{}^r_R(E, \mathbf{k}_{||}) = \begin{bmatrix} 0 & \cdots & 0 & & 0 \\ \vdots & & & & \vdots \\ 0 & \cdots & 0 & & 0 \\ 0 & \cdots & 0 & \sum_R^r(\zeta_{m+1}; E, \mathbf{k}_{||}) \end{bmatrix}$$
(9.67)

with

$$\Sigma_L^r(\zeta_0; E, \boldsymbol{k}_{||}) = B(\zeta_{-1})^\dagger \mathcal{G}_L^r(\zeta_{-1}, \zeta_{-1}; E, \boldsymbol{k}_{||}) B(\zeta_{-1})$$
$$\Sigma_R^r(\zeta_{m+1}; E, \boldsymbol{k}_{||}) = B(\zeta_{m+1}) \mathcal{G}_R^r(\zeta_{m+2}, \zeta_{m+2}; E, \boldsymbol{k}_{||}) B(\zeta_{m+1})^\dagger, \quad (9.68)$$

where $\mathcal{G}_{\{L,R\}}^r(\zeta_k, \zeta_l; E, \boldsymbol{k}_{||})$ is the N-dimensional (k, l) block-matrix element of $\hat{\mathcal{G}}_{\{L,R\}}^r(E, \boldsymbol{k}_{||})$. $\mathcal{G}_L^r(\zeta_{-1}, \zeta_{-1}; E, \boldsymbol{k}_{||})$ and $\mathcal{G}_R^r(\zeta_{m+2}, \zeta_{m+2}; E, \boldsymbol{k}_{||})$ are the Green's functions evaluated on the surfaces of the semi-infinite electrodes, which we refer to as the retarded surface Green's functions.

Theorem 9.1 states that the Green's function $G(\zeta_l, \zeta_m; Z \neq real, \boldsymbol{k}_{||})$ of a crystalline bulk is a decaying function (i.e., $G \to 0$) as $|l| \to \infty$ with m fixed (or as $|m| \to \infty$ with l fixed). Making use of this asymptotic property of the Green's function for nonreal Z, and employing Theorem 6.5 (see Section 6.5), we can prove the following theorem which gives a definite description of the surface Green's functions (or self-energy terms) in terms of the ratio matrices of the generalized Bloch states, R^{ref} and R^{tra}, introduced in Chapters 6 and 7. (For the efficient and stable calculation methods for R^{ref} and R^{tra}, see Sections 6.4 and 7.1.)

Theorem 9.2 *The retarded surface Green's functions of the left and right electrodes are explicitly expressed as*

$$\mathcal{G}_L^r(\zeta_{-1}, \zeta_{-1}; E, \boldsymbol{k}_{||}) = R^{ref}(\zeta_0; E, \boldsymbol{k}_{||}) B(\zeta_{-1})^{-1}$$
$$\mathcal{G}_R^r(\zeta_{m+2}, \zeta_{m+2}; E, \boldsymbol{k}_{||}) = R^{tra}(\zeta_{m+2}; E, \boldsymbol{k}_{||}) B(\zeta_{m+1})^{\dagger-1}, \quad (9.69)$$

and then, the retarded self-energy terms (9.68) are given by

$$\Sigma_L^r(\zeta_0; E, \boldsymbol{k}_{||}) = B(\zeta_{-1})^\dagger R^{ref}(\zeta_0; E, \boldsymbol{k}_{||})$$
$$\Sigma_R^r(\zeta_{m+1}; E, \boldsymbol{k}_{||}) = B(\zeta_{m+1}) R^{tra}(\zeta_{m+2}; E, \boldsymbol{k}_{||}). \quad (9.70)$$

Proof. We verify the expression of $\mathcal{G}_L^r$ in (9.69), mainly following the procedure of Lee and Joannopoulos (1981b); the expression of $\mathcal{G}_R^r$ is similarly obtained. We already know in Theorem 6.5 that the eigenvalues λ_n's of the generalized eigenvalue equation (6.85) for nonreal Z are divided *evenly* into two groups with $|\lambda_n| > 1$ and $|\lambda_n| < 1$. This is also the case for the generalized eigenvalue equation

$$\Pi_1(Z, \boldsymbol{k}_{||}) \begin{bmatrix} \Phi_n(\zeta_m^{M-1}, Z, \boldsymbol{k}_{||}) \\ \Phi_n(\zeta_1^{M+1}, Z, \boldsymbol{k}_{||}) \end{bmatrix} = \lambda_n(Z, \boldsymbol{k}_{||}) \Pi_2(Z, \boldsymbol{k}_{||}) \begin{bmatrix} \Phi_n(\zeta_m^{M-1}, Z, \boldsymbol{k}_{||}) \\ \Phi_n(\zeta_1^{M+1}, Z, \boldsymbol{k}_{||}) \end{bmatrix}$$
$$(9.71)$$

which is the analytically continued equation of (7.1). Let $\{\Phi_n(\zeta_l; Z, \boldsymbol{k}_{||})\}$, $n = 1, 2, ..., N$, be a set of the N eigenstates of (9.71) for nonreal Z with

$|\lambda_n| > 1$ and $Q(\zeta_l; Z, \boldsymbol{k}_{||})$ be an N-dimensional matrix which gathers these eigenstates, i.e.,

$$Q(\zeta_l; Z, \boldsymbol{k}_{||}) = \left[\Phi_1(\zeta_l; Z, \boldsymbol{k}_{||}), \Phi_2(\zeta_l; Z, \boldsymbol{k}_{||}), ..., \Phi_N(\zeta_l; Z, \boldsymbol{k}_{||})\right]. \quad (9.72)$$

Thus, $\{\Phi_n(\zeta_l; Z \neq \text{real}, \boldsymbol{k}_{||})\}$ is the set of the N independent functions with respect to ζ_l with a decaying property such that

$$\Phi_n(\zeta_{l-pL}; Z, \boldsymbol{k}_{||}) = (\lambda_n)^{-p}\Phi_n(\zeta_l; Z, \boldsymbol{k}_{||}). \quad (9.73)$$

Here, L is the integer associated with the length of periodicity in the z direction, i.e., L_z/h_z, and p is an arbitrary positive integer. Furthermore, $\{\Phi_n(\zeta_l; Z, \boldsymbol{k}_{||})\}$ is the set of the N solutions of the Kohn–Sham equation with E replaced by a complex number Z,

$$-B_{l-2}^{\dagger}\Phi_n(\zeta_{l-2}; Z, \boldsymbol{k}_{||}) + A_{l-1}\Phi_n(\zeta_{l-1}; Z, \boldsymbol{k}_{||}) - B_{l-1}\Phi_n(\zeta_l; Z, \boldsymbol{k}_{||}) = 0, \quad (9.74)$$

where

$$A_l = Z - H(\zeta_l; \boldsymbol{k}_{||}) \quad \text{and} \quad B_l = B(\zeta_l). \quad (9.75)$$

We first derive the following expression of $\mathcal{G}_L(\zeta_l, \zeta_{-1}; Z, \boldsymbol{k}_{||})$ in terms of $Q(\zeta_l; Z, \boldsymbol{k}_{||})$: For $l = -1, -2, ...,$

$$\mathcal{G}_L(\zeta_l, \zeta_{-1}; Z, \boldsymbol{k}_{||}) = Q(\zeta_l; Z, \boldsymbol{k}_{||})Q(\zeta_0; Z, \boldsymbol{k}_{||})^{-1}(B_{-1})^{-1}. \quad (9.76)$$

By definition, $\{\mathcal{G}_L(\zeta_l, \zeta_{-1}; Z, \boldsymbol{k}_{||})\}$ satisfies

$$\begin{bmatrix} \ddots & \ddots & \ddots & & 0 \\ & -B_{-4}^{\dagger} & A_{-3} & -B_{-3} & \\ & & -B_{-3}^{\dagger} & A_{-2} & -B_{-2} \\ 0 & & & -B_{-2}^{\dagger} & A_{-1} \end{bmatrix} \begin{bmatrix} \vdots \\ \mathcal{G}_{-3,-1} \\ \mathcal{G}_{-2,-1} \\ \mathcal{G}_{-1,-1} \end{bmatrix} = \begin{bmatrix} \vdots \\ 0 \\ 0 \\ I \end{bmatrix}, \quad (9.77)$$

that is,

$$-B_{-2}^{\dagger}\mathcal{G}_{-2,-1} + A_{-1}\mathcal{G}_{-1,-1} = I \quad (9.78)$$

and

$$\left.\begin{array}{l} -B_{-3}^{\dagger}\mathcal{G}_{-3,-1} + A_{-2}\mathcal{G}_{-2,-1} - B_{-2}\mathcal{G}_{-1,-1} = 0 \\ -B_{-4}^{\dagger}\mathcal{G}_{-4,-1} + A_{-3}\mathcal{G}_{-3,-1} - B_{-3}\mathcal{G}_{-2,-1} = 0 \\ \vdots \end{array}\right\} \quad (9.79)$$

where

$$\mathcal{G}_{l,-1} = \mathcal{G}_L(\zeta_l, \zeta_{-1}; Z, \boldsymbol{k}_{||}). \quad (9.80)$$

By the ansatz that the N-dimensional $\mathcal{G}_L(\zeta_l, \zeta_{-1}; Z \neq \text{real}, \boldsymbol{k}_{||})$ decays deep inside the left electrode ($l \rightarrow -\infty$) and the linear independence of the decaying sequences of $\{\Phi_n(\zeta_l; Z, \boldsymbol{k}_{||})\}$, where $n = 1, 2, ..., N$, we can expand $\mathcal{G}_L$ in terms of $\{\Phi_n\}$, since (9.79) are the same sets of simultaneous linear equations as (9.74) for $l \leq -1$. [Similarly, $\mathcal{G}_R$ in the right electrode can be expressed as a linear combination of the N eigenstates with $|\lambda_n| < 1$ of (9.71).] Thus we have, for $l = -1, -2, ...,$

$$
\mathcal{G}_L(\zeta_l, \zeta_{-1}; Z, \boldsymbol{k}_{||}) = \left[\sum_{n=1}^{N} c_{n1}\Phi_n(\zeta_l; Z, \boldsymbol{k}_{||}), \sum_{n=1}^{N} c_{n2}\Phi_n(\zeta_l; Z, \boldsymbol{k}_{||}), \right.
$$
$$
\left. ..., \sum_{n=1}^{N} c_{nN}\Phi_n(\zeta_l; Z, \boldsymbol{k}_{||}) \right]
$$
$$
= Q(\zeta_l; Z, \boldsymbol{k}_{||})C \tag{9.81}
$$

with $\{c_{nn'}\}$ being a set of unknown expansion coefficients forming a matrix C. The dependence of $c_{nn'}$ and C on Z and $\boldsymbol{k}_{||}$ is ignored for simplicity. The coefficient matrix C is determined as

$$
C = Q(\zeta_0; Z, \boldsymbol{k}_{||})^{-1}(B_{-1})^{-1} \tag{9.82}
$$

by the insertion of (9.81) for $l = -1$ and -2 into (9.78) and the subsequent use of (9.74) for $l = 0$. Equation (9.76) is now obtained.

Next, we carry out the limiting procedure in (9.76) to give the retarded Green's function as

$$
\mathcal{G}_L^r(\zeta_l, \zeta_{-1}; E, \boldsymbol{k}_{||}) = \lim_{\eta \rightarrow 0+} \mathcal{G}_L(\zeta_l, \zeta_{-1}; E + i\eta, \boldsymbol{k}_{||})
$$
$$
= Q^{ref}(\zeta_l; E, \boldsymbol{k}_{||})Q^{ref}(\zeta_0; E, \boldsymbol{k}_{||})^{-1}(B_{-1})^{-1}, \tag{9.83}
$$

since (6.105) implies that

$$
\lim_{\eta \rightarrow 0+} Q(\zeta_l; E + i\eta, \boldsymbol{k}_{||}) = Q^{ref}(\zeta_l; E, \boldsymbol{k}_{||}). \tag{9.84}
$$

Finally, we obtain the expression of $\mathcal{G}_L^r$ in (9.69) from (9.83) for $l = -1$ by using the definition of the ratio matrix R^{ref}, (7.2).

In addition, we give another intuitive derivation of (9.76) for $l = 1$, and therefore, (9.69). The continued-fraction equation, analogous to those of Theorem 6.4 and (7.3),

$$
R(\zeta_0; Z, \boldsymbol{k}_{||}) = \mathcal{G}_{-1,-1}^p B_{-1} + \mathcal{G}_{-1,-pm}^p B_{-1}^\dagger
$$
$$
\times \left[R(\zeta_0; Z, \boldsymbol{k}_{||})^{-1} - \mathcal{G}_{-pm,-pm}^p B_{-1}^\dagger \right]^{-1} \mathcal{G}_{-pm,-1}^p B_{-1} \tag{9.85}
$$

is satisfied by the ratio matrix $R(\zeta_0 : Z, \mathbf{k}_{||})$ which is defined as

$$R(\zeta_0 : Z, \mathbf{k}_{||}) = Q(\zeta_{-1}; Z, \mathbf{k}_{||})Q(\zeta_0; Z, \mathbf{k}_{||})^{-1}. \qquad (9.86)$$

Here,

$$\mathcal{G}^p_{k,l} = \mathcal{G}^p(\zeta_k, \zeta_l; Z, \mathbf{k}_{||}) \qquad (9.87)$$

is the (k, l) block-matrix element of the Green's function of the truncated region overlying p unit cells in the left electrode and m is the number of ζ-grid points within one unit cell. When $Z \neq$ real, as $p \to \infty$ in (9.85), the off-diagonal $\mathcal{G}^p_{-1,-pm}$ and $\mathcal{G}^p_{-pm,-1}$ go to zero and the diagonal $\mathcal{G}^p_{-1,-1}$ becomes the surface Green's function $\mathcal{G}_L(\zeta_{-1}, \zeta_{-1}; Z, \mathbf{k}_{||})$. $\qquad \square$

The representation of the surface Green's functions by generalized Bloch states has been investigated by some authors [Lee and Joannopoulos (1981b), Sanvito *et al.* (1999) and Taylor *et al.* (2001)]. It should be noted that Theorem 9.2 is particularly attractive because this theorem allows us to directly evaluate the retarded surface Green's functions [or retarded self-energy terms (9.68)] at a purely real E without relying on a finite broadening (or smearing) parameter η which is employed in most iterative methods. For iterative methods using a small parameter η ($\neq 0$), see, for example, Guinea *et al.* (1983), Lópes Sancho *et al.* (1984, 1985), Martín-Moreno and Vergés (1990), Cerdá *et al.* (1997), Buongiorno Nardelli (1999), Palacios *et al.* (2002) and Thygesen *et al.* (2003).

9.4 Green's Function Matching in the Overbridging Boundary-Matching Scheme and Related Topics

We here focus on the retarded Green's function of the whole system portioned to the transition region, which is derived from (9.60),

$$\hat{G}^r_T(E, \mathbf{k}_{||}) = \lim_{\eta \to 0^+} \hat{G}_T(E + i\eta, \mathbf{k}_{||})$$

$$= \left[E - \hat{H}_T(\mathbf{k}_{||}) - \hat{\Sigma}^r_L(E, \mathbf{k}_{||}) - \hat{\Sigma}^r_R(E, \mathbf{k}_{||}) \right]^{-1} \qquad (9.88)$$

in the general case of the Hamiltonian matrix $\hat{H}_T(\mathbf{k}_{||})$ being not necessarily sparse. The simplest way to calculate $\hat{G}^r_T(E, \mathbf{k}_{||})$ of (9.88) might be to carry out direct matrix inversion. In practice, however, it is a computationally hard task to perform inversion calculations with large-sized matrices. Moreover, standard iterative algorithms, such as the steepest-descent (SD) and conjugate-gradient (CG) algorithms, are not always effective in the present case, since the matrix in (9.88) is non-Hermitian, for which pathologically slow convergence is frequently a serious problem. Nevertheless, in what follows, we show that there exists an efficient approach to computing the non-Hermitian $\hat{G}_T^r(E, \boldsymbol{k}_{||})$ based on the overbridging boundary-matching (OBM) scheme.

Let us consider the l-th column of $\hat{G}_T^r(E, \boldsymbol{k}_{||})$, i.e., $[G_T^r(\zeta_0, \zeta_l; E, \boldsymbol{k}_{||}),$ $G_T^r(\zeta_1, \zeta_l; E, \boldsymbol{k}_{||}), ..., G_T^r(\zeta_{m+1}, \zeta_l; E, \boldsymbol{k}_{||})]^t$ $(l = 0, 1, 2, ..., m + 1)$. From (9.88), one easily finds that the l-th column satisfies the following equation by virtue of a simple form of the self-energy matrices (9.67):

$$
\left[E - \hat{H}_T(\boldsymbol{k}_{||}) \right]
\begin{bmatrix}
G_T^r(\zeta_0, \zeta_l) \\
G_T^r(\zeta_1, \zeta_l) \\
\vdots \\
G_T^r(\zeta_l, \zeta_l) \\
\vdots \\
G_T^r(\zeta_m, \zeta_l) \\
G_T^r(\zeta_{m+1}, \zeta_l)
\end{bmatrix}
=
\begin{bmatrix}
\Sigma_L^r(\zeta_0) G_T^r(\zeta_0, \zeta_l) \\
0 \\
\vdots \\
0 \\
I \\
0 \\
\vdots \\
0 \\
\Sigma_R^r(\zeta_{m+1}) G_T^r(\zeta_{m+1}, \zeta_l)
\end{bmatrix}
, \leftarrow \text{the } l\text{-th}
$$

$$(9.89)$$

where

$$
\begin{aligned}
G_T^r(\zeta_k, \zeta_l) &= G_T^r(\zeta_k, \zeta_l; E, \boldsymbol{k}_{||}) \\
\Sigma_{\{L,R\}}^r(\zeta_k) &= \Sigma_{\{L,R\}}^r(\zeta_k; E, \boldsymbol{k}_{||}).
\end{aligned}
\tag{9.90}
$$

Equation (9.89) provides a matching relation with regard to the retarded Green's function $\{G_T^r(\zeta_k, \zeta_l)\}$, which should be compared with (6.9), an analogous matching relation with regard to the wave function $\{\Psi(z_k)\}$. From (9.89), one can see that once the unperturbed Green's function $\hat{\mathcal{G}}_T(E, \boldsymbol{k}_{||}) = [E - \hat{H}_T(\boldsymbol{k}_{||})]^{-1}$ is known, the perturbed Green's function elements $G_T^r(\zeta_0, \zeta_l)$ and $G_T^r(\zeta_{m+1}, \zeta_l)$ are calculated by

$$
\begin{bmatrix}
\mathcal{G}_T(\zeta_0, \zeta_0)\Sigma_L^r(\zeta_0) - I & \mathcal{G}_T(\zeta_0, \zeta_{m+1})\Sigma_R^r(\zeta_{m+1}) \\
\mathcal{G}_T(\zeta_{m+1}, \zeta_0)\Sigma_L^r(\zeta_0) & \mathcal{G}_T(\zeta_{m+1}, \zeta_{m+1})\Sigma_R^r(\zeta_{m+1}) - I
\end{bmatrix}
\begin{bmatrix}
G_T^r(\zeta_0, \zeta_l) \\
G_T^r(\zeta_{m+1}, \zeta_l)
\end{bmatrix}
$$

$$
= -
\begin{bmatrix}
\mathcal{G}_T(\zeta_0, \zeta_l) \\
\mathcal{G}_T(\zeta_{m+1}, \zeta_l)
\end{bmatrix}. \tag{9.91}
$$

Here,

$$
\mathcal{G}_T(\zeta_k, \zeta_l) = \mathcal{G}_T(\zeta_k, \zeta_l; E, \boldsymbol{k}_{||})
\tag{9.92}
$$

is the N-dimensional (k, l) block-matrix element of $\hat{\mathcal{G}}_T(E, \boldsymbol{k}_{||})$. Equation (9.91), which is $2N$ simultaneous linear equations with respect to

$G_T^r(\zeta_0, \zeta_l)$ and $G_T^r(\zeta_{m+1}, \zeta_l)$, manifests boundary-value (surface) matching for the retarded Green's function of the whole system. The surface Green's function matching theory has been pioneered by García-Moliner *et al.* See Barrio *et al.* (1989), García-Moliner *et al.* (1990) and García-Moliner and Velasco (1991, 1992).

Recall that making efficient use of the SD and CG algorithms, we can calculate, with rapid convergence, the l-th column of $\hat{\mathcal{G}}_T(E, \boldsymbol{k}_{||})$ ($=$ $[E - \hat{H}_T(\boldsymbol{k}_{||})]^{-1}$), i.e., $[\mathcal{G}_T(\zeta_0, \zeta_l), \mathcal{G}_T(\zeta_1, \zeta_l), ..., \mathcal{G}_T(\zeta_{m+1}, \zeta_l)]^t$, thanks to the Hermiticity of the Hamiltonian $\hat{H}_T(\boldsymbol{k}_{||})$, which was argued in relation to wave-function matching in Section 6.2. The computational cost of the subsequent calculation is now greatly reduced because the dimension of the relevant non-Hermitian matrix is downsized from $(m+2)N$ of (9.88) to $2N$ of (9.91); the solution of (9.91) is obtainable by direct matrix inversion with a moderate computational effort, and in the particular cases of $l = 0$ and $m + 1$, the following analytic expressions for the solution of (9.91) are available. For $l = 0$,

$$G_T^r(\zeta_0, \zeta_0) = \tilde{\mathcal{G}}_T(\zeta_0, \zeta_0) \left[I - \textstyle\sum_L^r(\zeta_0) \tilde{\mathcal{G}}_T(\zeta_0, \zeta_0) \right]^{-1}$$

$$G_T^r(\zeta_{m+1}, \zeta_0) = \left[I - \mathcal{G}_T(\zeta_{m+1}, \zeta_{m+1}) \textstyle\sum_R^r(\zeta_{m+1}) \right]^{-1}$$
$$\times \mathcal{G}_T(\zeta_{m+1}, \zeta_0) \left[I - \textstyle\sum_L^r(\zeta_0) \tilde{\mathcal{G}}_T(\zeta_0, \zeta_0) \right]^{-1}, \quad (9.93)$$

where $\tilde{\mathcal{G}}_T(\zeta_0, \zeta_0)$ is a modified $\mathcal{G}_T(\zeta_0, \zeta_0)$ under the influence of the right-electrode self-energy term $\textstyle\sum_R^r(\zeta_{m+1})$, which is expressed as

$$\tilde{\mathcal{G}}_T(\zeta_0, \zeta_0) = \mathcal{G}_T(\zeta_0, \zeta_0)$$
$$+ \mathcal{G}_T(\zeta_0, \zeta_{m+1}) \textstyle\sum_R^r(\zeta_{m+1}) \left[I - \mathcal{G}_T(\zeta_{m+1}, \zeta_{m+1}) \textstyle\sum_R^r(\zeta_{m+1}) \right]^{-1}$$
$$\times \mathcal{G}_T(\zeta_{m+1}, \zeta_0). \quad (9.94)$$

For $l = m + 1$,

$$G_T^r(\zeta_0, \zeta_{m+1}) = \left[I - \mathcal{G}_T(\zeta_0, \zeta_0) \textstyle\sum_L^r(\zeta_0) \right]^{-1} \mathcal{G}_T(\zeta_0, \zeta_{m+1})$$
$$\times \left[I - \textstyle\sum_R^r(\zeta_{m+1}) \tilde{\mathcal{G}}_T(\zeta_{m+1}, \zeta_{m+1}) \right]^{-1}$$

$$G_T^r(\zeta_{m+1}, \zeta_{m+1}) = \tilde{\mathcal{G}}_T(\zeta_{m+1}, \zeta_{m+1}) \left[I - \textstyle\sum_R^r(\zeta_{m+1}) \tilde{\mathcal{G}}_T(\zeta_{m+1}, \zeta_{m+1}) \right]^{-1}$$
$$(9.95)$$

with $\tilde{\mathcal{G}}_T(\zeta_{m+1}, \zeta_{m+1})$ being a similar modified $\mathcal{G}_T(\zeta_{m+1}, \zeta_{m+1})$ that is in turn affected by the left-electrode self-energy term $\sum_L^r(\zeta_0)$, i.e.,

$$\tilde{\mathcal{G}}_T(\zeta_{m+1}, \zeta_{m+1}) = \mathcal{G}_T(\zeta_{m+1}, \zeta_{m+1})$$
$$+ \mathcal{G}_T(\zeta_{m+1}, \zeta_0) \sum_L^r(\zeta_0) \Big[I - \mathcal{G}_T(\zeta_0, \zeta_0) \sum_L^r(\zeta_0) \Big]^{-1}$$
$$\times \mathcal{G}_T(\zeta_0, \zeta_{m+1}). \tag{9.96}$$

It is easy to make sure that G_T^r's given by (9.93) – (9.96) satisfy (9.91), and therefore, they are exact analytic solutions of Dyson's equation (9.64) associated with the retarded Green's function.

We next exhibit that interesting relationships between the retarded Green's functions and the scattering wave functions emerge upon comparison of the Green's function matching formula (9.91) and the wave-function matching formula (7.6) [or (6.26)]. The latter is rewritten in terms of $\mathcal{G}_T(\zeta_k, \zeta_l)$ of (7.8), the self-energy matrices $\sum_{\{L,R\}}^r$ of (9.70), and the coupling matrix Γ_L as

$$\begin{bmatrix} \mathcal{G}_T(\zeta_0, \zeta_0) \sum_L^r(\zeta_0) - I & \mathcal{G}_T(\zeta_0, \zeta_{m+1}) \sum_R^r(\zeta_{m+1}) \\ \mathcal{G}_T(\zeta_{m+1}, \zeta_0) \sum_L^r(\zeta_0) & \mathcal{G}_T(\zeta_{m+1}, \zeta_{m+1}) \sum_R^r(\zeta_{m+1}) - I \end{bmatrix} \begin{bmatrix} \Psi_L(\zeta_0) \\ \Psi_L(\zeta_{m+1}) \end{bmatrix}$$
$$= -i \begin{bmatrix} \mathcal{G}_T(\zeta_0, \zeta_0) \\ \mathcal{G}_T(\zeta_{m+1}, \zeta_0) \end{bmatrix} \Gamma_L(\zeta_0) \Phi_L^{in}(\zeta_0). \tag{9.97}$$

Here, the coupling matrix $\Gamma_L(\zeta_0)$ that describes the 'coupling strength' of the transition region to the left electrode at ζ_0 (more concretely, corresponding to the flux or the group velocity, as seen later) is defined by

$$\Gamma_L(\zeta_0) = i \Big[\sum_L^r(\zeta_0) - \sum_L^a(\zeta_0) \Big] \qquad \Big(\sum_L^a(\zeta_0) = \sum_L^r(\zeta_0)^\dagger \Big), \tag{9.98}$$

and scattering wave functions Ψ_L are marked with the subscript L as a reminder that an incident wave Φ_L^{in} incoming from deep inside the *left* electrode is relevant to the present case. In the derivation of the right-hand side of (9.97) from that of (7.6), we introduced the following ratio matrix R^{in} in the left electrode ($l \leq 0$) along a similar line to the definition of R^{ref} of (7.2):

$$R^{in}(\zeta_l) = Q^{in}(\zeta_{l-1}) Q^{in}(\zeta_l)^{-1}, \tag{9.99}$$

where

$$Q^{in}(\zeta_l) = \Big[\Phi_{L,1}^{in}(\zeta_l), \Phi_{L,2}^{in}(\zeta_l), ..., \Phi_{L,N}^{in}(\zeta_l) \Big], \tag{9.100}$$

which is assumed to include not only ordinary right-propagating incident Bloch waves but also leftward decreasing evanescent waves. Furthermore, we used

$$\Phi_L^{in}(\zeta_{-1}) = R^{in}(\zeta_0)\Phi_L^{in}(\zeta_0) \tag{9.101}$$

which comes from the definition of $R^{in}(\zeta_0)$, (9.99), and employed the relationship between the advanced self-energy term $\sum_L^a(\zeta_0)$ and $R^{in}(\zeta_0)$:

$$\sum_L^a(\zeta_0) = B(\zeta_{-1})^\dagger R^{in}(\zeta_0), \tag{9.102}$$

[Equation (9.102) can be verified by the same argument as the proof of (9.70). In this case, however, a limiting procedure such that

$$\lim_{\eta \to 0^+} Q(\zeta_l; E - i\eta, \mathbf{k}_{||}) = Q^{in}(\zeta_l; E, \mathbf{k}_{||}) \tag{9.103}$$

is implemented instead of (9.84).] Now, noting the similarity between (9.91) for $l = 0$ and (9.97), we immediately have the relations in question:

$$\Psi_L(\zeta_0) = iG_T^r(\zeta_0, \zeta_0)\Gamma_L(\zeta_0)\Phi_L^{in}(\zeta_0)$$
$$\Psi_L(\zeta_{m+1}) = iG_T^r(\zeta_{m+1}, \zeta_0)\Gamma_L(\zeta_0)\Phi_L^{in}(\zeta_0). \tag{9.104}$$

We also obtain the corresponding relations in the case of an incident wave Φ_R^{in} incoming from the *right* electrode:

$$\Psi_R(\zeta_0) = iG_T^r(\zeta_0, \zeta_{m+1})\Gamma_R(\zeta_{m+1})\Phi_R^{in}(\zeta_{m+1})$$
$$\Psi_R(\zeta_{m+1}) = iG_T^r(\zeta_{m+1}, \zeta_{m+1})\Gamma_R(\zeta_{m+1})\Phi_R^{in}(\zeta_{m+1}), \tag{9.105}$$

where

$$\Gamma_R(\zeta_{m+1}) = i\left[\sum_R^r(\zeta_{m+1}) - \sum_R^a(\zeta_{m+1})\right] \quad \left(\sum_R^a(\zeta_{m+1}) = \sum_R^r(\zeta_{m+1})^\dagger\right). \tag{9.106}$$

Analogous relationships between the Green's functions and the scattering wave functions have been discussed by Barrio *et al.* (1989), García-Moliner *et al.* (1990), García-Moliner and Velasco (1991, 1992), Cerdá *et al.* (1997) and Sanvito *et al.* (1999).

We next show that the expectation value $\Phi_L^{in}(\zeta_0)^\dagger \Gamma_L(\zeta_0)\Phi_L^{in}(\zeta_0)$ $\left(\Phi_R^{in}(\zeta_{m+1})^\dagger \Gamma_R(\zeta_{m+1})\Phi_R^{in}(\zeta_{m+1})\right)$ is proportional to the flux through the x–y plane at ζ_0 (ζ_{m+1}), or the z component of the group velocity, of the state $\Phi_L^{in}(\zeta_0)$ ($\Phi_R^{in}(\zeta_{m+1})$). This is because one obtains, from (9.98), (9.101) and (9.102),

$$\Phi_L^{in}(\zeta_0)^\dagger \Gamma_L(\zeta_0)\Phi_L^{in}(\zeta_0) = i\left[\Phi_L^{in}(\zeta_{-1})^\dagger B(\zeta_{-1})\Phi_L^{in}(\zeta_0)\right.$$
$$\left. - \Phi_L^{in}(\zeta_0)^\dagger B(\zeta_{-1})^\dagger \Phi_L^{in}(\zeta_{-1})\right] \tag{9.107}$$

which is similar to (6.113) for $k = -1$, namely, the same as the definition of the flux $\mathcal{I}(\zeta_{-1})$ for a state normalized to unity over a unit cell. [This is also the case for $\Phi_R^{in}(\zeta_{m+1})^\dagger \Gamma_R(\zeta_{m+1}) \Phi_R^{in}(\zeta_{m+1}).$] Therefore, it is concluded that the coupling matrices $\Gamma_L(\zeta_0)$ and $\Gamma_R(\zeta_{m+1})$ are Hermitian matrices which have a definite physical meaning of the flux.

Finally, we address the problem of electronic transport within the framework of the Green's function approach. Using (9.104) and (9.105), we can prove that the Landauer–Büttiker formula (6.128) describing the conductance G in the zero bias limit has the following expression in terms of the Green's functions $G_T^{\{r,a\}}$ and the coupling matrices $\Gamma_{\{L,R\}}$, where all of the quantities are evaluated at the Fermi level E_F. One of the advantages of the Green's function approach is that the conductance is calculated without the knowledge of well-defined asymptotic wave functions in conducting channels.

Theorem 9.3

$$
\begin{aligned}
G &= \frac{2e^2}{h} \sum_{i,j} |t_{i,j}|^2 \frac{v_i'}{v_j} \\
&= \frac{2e^2}{h} Tr\Big[\Gamma_L(\zeta_0) G_T^a(\zeta_0, \zeta_{m+1}) \Gamma_R(\zeta_{m+1}) G_T^r(\zeta_{m+1}, \zeta_0) \Big] \\
&= \frac{2e^2}{h} Tr\Big[\Gamma_L(\zeta_0) G_T^r(\zeta_0, \zeta_{m+1}) \Gamma_R(\zeta_{m+1}) G_T^a(\zeta_{m+1}, \zeta_0) \Big] \quad (9.108)
\end{aligned}
$$

Here, 'Tr' represents the trace, i.e., the sum of the diagonal matrix elements, and

$$
G_T^a(\zeta_k, \zeta_l) = G_T^r(\zeta_l, \zeta_k)^\dagger. \quad (9.109)
$$

Proof. Let us consider the case of incident waves incoming from deep inside the left electrode. The scattering wave function $\Psi_{L,j}(\zeta_l)$ corresponding to the j-th incident wave $\Phi_{L,j}^{in}(\zeta_0)$ is given by a linear combination of transmitted waves $\Phi_{L,i}^{tra}(\zeta_l)$ inside the right electrode ($l \geq m+1$) with transmission coefficient $t_{i,j}$, i.e.,

$$
\Psi_{L,j}(\zeta_l) = \sum_{i=1}^{N} t_{i,j} \Phi_{L,i}^{tra}(\zeta_l) = Q^{tra}(\zeta_l) \begin{bmatrix} t_{1j} \\ t_{2j} \\ \vdots \\ t_{Nj} \end{bmatrix}, \quad (9.110)
$$

where $Q^{tra}(\zeta_l)$ is defined by (7.5), i.e.,

$$
Q^{tra}(\zeta_l) = \Big[\Phi_{L,1}^{tra}(\zeta_l), \Phi_{L,2}^{tra}(\zeta_l), ..., \Phi_{L,N}^{tra}(\zeta_l) \Big]. \quad (9.111)
$$

From (9.104) and (9.110) for $l = m + 1$, we obtain

$$
\begin{bmatrix} t_{1j} \\ t_{2j} \\ \vdots \\ t_{Nj} \end{bmatrix} = iQ^{tra}(\zeta_{m+1})^{-1}G_T^r(\zeta_{m+1}, \zeta_0)\Gamma_L(\zeta_0)\Phi_{L,j}^{in}(\zeta_0) \tag{9.112}
$$

and then have

$$
T = iQ^{tra}(\zeta_{m+1})^{-1}G_T^r(\zeta_{m+1}, \zeta_0)\Gamma_L(\zeta_0)Q^{in}(\zeta_0). \tag{9.113}
$$

Here, T is the transmission-coefficient matrix

$$
T = \begin{bmatrix} t_{11} & t_{12} & \cdots & t_{1N} \\ t_{21} & t_{22} & \cdots & t_{2N} \\ & \cdots & \\ t_{N1} & t_{N2} & \cdots & t_{NN} \end{bmatrix} \tag{9.114}
$$

and $Q^{in}(\zeta_0)$ is the matrix defined by (9.100). Noting that the expression

$$
\sum_{i,j} |t_{ij}|^2 \frac{v_i'}{v_j} = Tr\left[\mathcal{V}^{-1}T^\dagger \mathcal{V}'T\right] \tag{9.115}
$$

holds with $\mathcal{V}^{(')}$ being a diagonal matrix whose elements are $v_i^{(')}\delta_{ij}$, and substituting (9.113) into the right-hand side of (9.115), we find

$$
\sum_{i,j} |t_{i,j}|^2 \frac{v_i'}{v_j} = Tr\Big[\Gamma_L(\zeta_0)Q^{in}(\zeta_0)\mathcal{V}^{-1}Q^{in}(\zeta_0)^\dagger\Gamma_L(\zeta_0)G_T^a(\zeta_0, \zeta_{m+1})
$$
$$
\times Q^{tra}(\zeta_{m+1})^{\dagger-1}\mathcal{V}'Q^{tra}(\zeta_{m+1})^{-1}G_T^r(\zeta_{m+1}, \zeta_0)\Big]. \tag{9.116}
$$

Here we used the cyclic property of the trace. Accordingly, it is seen that the proof of the second line in (9.108) is accomplished, provided that the relations

$$
\mathcal{V} = \alpha Q^{in}(\zeta_0)^\dagger\Gamma_L(\zeta_0)Q^{in}(\zeta_0)
$$
$$
\mathcal{V}' = \alpha Q^{tra}(\zeta_{m+1})^\dagger\Gamma_R(\zeta_{m+1})Q^{tra}(\zeta_{m+1}) \tag{9.117}
$$

are established, where $\alpha \ (\neq 0)$ is a certain constant. In fact, we have the following expressions for the velocity matrices $\mathcal{V}$ and $\mathcal{V}'$:

$$\mathcal{V} = iL_z\left[Q^{in}(\zeta_{-1})^\dagger B(\zeta_{-1})Q^{in}(\zeta_0) - Q^{in}(\zeta_0)^\dagger B(\zeta_{-1})^\dagger Q^{in}(\zeta_{-1})\right]$$

$$= L_z Q^{in}(\zeta_0)^\dagger \Gamma_L(\zeta_0)Q^{in}(\zeta_0)$$

$$\mathcal{V}' = iL_z\left[Q^{tra}(\zeta_m)^\dagger B(\zeta_m)Q^{tra}(\zeta_{m+1}) - Q^{tra}(\zeta_{m+1})^\dagger B(\zeta_m)^\dagger Q^{tra}(\zeta_m)\right]$$

$$= L_z Q^{tra}(\zeta_{m+1})^\dagger \Gamma_R(\zeta_{m+1})Q^{tra}(\zeta_{m+1}), \tag{9.118}$$

where L_z is the unit-cell length in the z direction and wave functions are assumed to be normalized in the unit cell. Equations (9.118) are derived from the representation of the group velocity, (6.121), using (6.117) and the relationship between the coupling matrix Γ_L and the flux $\mathcal{I}$, (9.107). The equality of the last line in (9.108) is also verified in a similar consideration of the case of incident waves incoming from the right electrode.   □

Equation (9.108) is a well-known formula derived from the nonequilibrium Green's function (NEGF) formalism pioneered by Keldysh (1964). There have been a number of works concerning NEGF and the application of (9.108), and we refer interested readers to the contributions. For example, see Caroli *et al.* (1971), Lipavský *et al.* (1986), Pernas *et al.* (1990), McLennan *et al.* (1991), Meir and Wingreen (1992), Levy Yeyati (1992), Todorov *et al.* (1993), Jauho *et al.* (1994), Cuevas *et al.* (1996, 1998), Brandbyge *et al.* (1999, 2002), Wang *et al.* (1999), Taylor *et al.* (2001), Buongiorno Nardelli *et al.* (2001), Nakanishi and Tsukada (2001), Larade *et al.* (2001), Damle *et al.* (2001), Palacios *et al.* (2002), Xue *et al.* (2002), Thygesen *et al.* (2003), Long *et al.* (2003), Thygesen and Jacobsen (2003), Larade and Bratkovsky (2003), Tamura (2003), Liu and Guo (2004), Liang *et al.* (2004). For a review, see Datta (1995), in which (9.108) is shown to be also derived from the Fisher–Lee relation [Fisher and Lee (1981)].

9.5   Green's Function of a Whole System with a Block-Tridiagonal Hamiltonian Matrix

In this section, we limit our discussion to the retarded Green's function of a whole system described by the Hamiltonian matrix of a block-tridiagonal form, whose representation in the transition region is

$$\hat{H}_T(\bm{k}_{||}) = \begin{bmatrix} H_0 & B_0 & & & & 0 \\ B_0^\dagger & H_1 & B_1 & & & \\ & B_1^\dagger & H_2 & B_2 & & \\ & & \ddots & \ddots & \ddots & \\ & & & B_{m-1}^\dagger & H_m & B_m \\ 0 & & & & B_m^\dagger & H_{m+1} \end{bmatrix}, \qquad (9.119)$$

where

$$H_l = H(\zeta_l; \bm{k}_{||}) \quad \text{and} \quad B_l = B(\zeta_l) \qquad (9.120)$$

which are N-dimensional block matrices (see Fig. 9.2).

For determining the electronic charge density in the Green's function formalism, it is necessary to compute the diagonal elements of the retarded Green's function matrix [see Eq. (9.15)]. The following theorem is useful for this purpose.

Theorem 9.4

$$G_T^r(\zeta_l, \zeta_l; E, \bm{k}_{||}) = \left[A_l - B_{l-1}^\dagger X_l - B_l Y_l \right]^{-1}$$
$$(0 \le l \le m+1) \qquad (9.121)$$

Here, X_l $(1 \le l \le m+1)$ and Y_l $(0 \le l \le m)$ are represented in continued-fraction forms as

$$X_l = \left[A_{l-1} - B_{l-2}^\dagger \left[A_{l-2} - \cdots - B_0^\dagger \left[A_0 - \Sigma_L^r(\zeta_0; E, \bm{k}_{||}) \right]^{-1} B_0 \right. \right.$$
$$\left. \left. \cdots \right]^{-1} B_{l-3} \right]^{-1} B_{l-2} \right]^{-1} B_{l-1}$$

$$Y_l = \left[A_{l+1} - B_{l+1} \left[A_{l+2} - \cdots - B_m \left[A_{m+1} - \Sigma_R^r(\zeta_{m+1}; E, \bm{k}_{||}) \right]^{-1} B_m^\dagger \right. \right.$$
$$\left. \left. \cdots \right]^{-1} B_{l+2}^\dagger \right]^{-1} B_{l+1}^\dagger \right]^{-1} B_l^\dagger \qquad (9.122)$$

with A_l being $E - H_l$, and

$$X_0 = (B_{-1}^\dagger)^{-1} \Sigma_L^r(\zeta_0; E, \bm{k}_{||})$$
$$Y_{m+1} = (B_{m+1})^{-1} \Sigma_R^r(\zeta_{m+1}; E, \bm{k}_{||}). \qquad (9.123)$$

Proof. The l-th column $[G_T^r(\zeta_0, \zeta_l), G_T^r(\zeta_1, \zeta_l), ..., G_T^r(\zeta_{m+1}, \zeta_l)]^t$
($l = 0, 1, ..., m+1$) satisfies (9.89); more definitely,

$$
\begin{bmatrix}
A_0 & -B_0 & & & & & 0 \\
-B_0^\dagger & A_1 & -B_1 & & & & \\
& \ddots & \ddots & \ddots & & & \\
& & -B_{l-1}^\dagger & A_l & -B_l & & \\
& & & \ddots & \ddots & \ddots & \\
& & & & -B_{m-1}^\dagger & A_m & -B_m \\
0 & & & & & -B_m^\dagger & A_{m+1}
\end{bmatrix}
\begin{bmatrix}
G_T^r(\zeta_0, \zeta_l) \\
G_T^r(\zeta_1, \zeta_l) \\
\vdots \\
G_T^r(\zeta_l, \zeta_l) \\
\vdots \\
G_T^r(\zeta_m, \zeta_l) \\
G_T^r(\zeta_{m+1}, \zeta_l)
\end{bmatrix}
$$

$$
=
\begin{bmatrix}
\Sigma_L^r(\zeta_0) G_T^r(\zeta_0, \zeta_l) \\
0 \\
\vdots \\
I \\
\vdots \\
0 \\
\Sigma_R^r(\zeta_{m+1}) G_T^r(\zeta_{m+1}, \zeta_l)
\end{bmatrix}
. \leftarrow \text{the } l\text{-th}
$$

$$(9.124)$$

For $l = 1, 2, ..., m$, we then have

$$
\left.
\begin{aligned}
(A_0 - \Sigma_L^r(\zeta_0))G_T^r(\zeta_0, \zeta_l) - B_0 G_T^r(\zeta_1, \zeta_l) &= 0 \\
-B_0^\dagger G_T^r(\zeta_0, \zeta_l) + A_1 G_T^r(\zeta_1, \zeta_l) - B_1 G_T^r(\zeta_2, \zeta_l) &= 0 \\
\vdots \\
-B_{l-2}^\dagger G_T^r(\zeta_{l-2}, \zeta_l) + A_{l-1} G_T^r(\zeta_{l-1}, \zeta_l) - B_{l-1} G_T^r(\zeta_l, \zeta_l) &= 0
\end{aligned}
\right\}
\quad (9.125)
$$

and

$$
-B_{l-1}^\dagger G_T^r(\zeta_{l-1}\zeta_l) + A_l G_T^r(\zeta_l, \zeta_l) - B_l G_T^r(\zeta_{l+1}, \zeta_l) = I \qquad (9.126)
$$

and

$$
\left.
\begin{aligned}
-B_l^\dagger G_T^r(\zeta_l, \zeta_l) + A_{l+1} G_T^r(\zeta_{l+1}, \zeta_l) - B_{l+1} G_T^r(\zeta_{l+2}, \zeta_l) &= 0 \\
\vdots \\
-B_{m-1}^\dagger G_T^r(\zeta_{m-1}, \zeta_l) + A_m G_T^r(\zeta_m, \zeta_l) - B_m G_T^r(\zeta_{m+1}, \zeta_l) &= 0 \\
-B_m^\dagger G_T^r(\zeta_m, \zeta_l) + (A_{m+1} - \Sigma_R^r(\zeta_{m+1}))G_T^r(\zeta_{m+1}, \zeta_l) &= 0
\end{aligned}
\right\}
. \quad (9.127)
$$

Equation (9.126) multiplied by $G_T^r(\zeta_l, \zeta_l)^{-1}$ from the right-hand side leads to

$$G_T^r(\zeta_l, \zeta_l) = \Big[A_l - B_{l-1}^\dagger G_T^r(\zeta_{l-1}, \zeta_l)G_T^r(\zeta_l, \zeta_l)^{-1}$$

$$-B_l G_T^r(\zeta_{l+1}, \zeta_l)G_T^r(\zeta_l, \zeta_l)^{-1}\Big]^{-1}, \tag{9.128}$$

which is seen to be equal to (9.121) if X_l and Y_l are defined by

$$X_l = G_T^r(\zeta_{l-1}, \zeta_l)G_T^r(\zeta_l, \zeta_l)^{-1}$$
$$Y_l = G_T^r(\zeta_{l+1}, \zeta_l)G_T^r(\zeta_l, \zeta_l)^{-1}. \tag{9.129}$$

Introducing ratio matrices associated with the retarded Green's function

$$F_{i,j}^{(k)} = G_T^r(\zeta_i, \zeta_k)G_T^r(\zeta_j, \zeta_k)^{-1}, \tag{9.130}$$

we have $X_l = F_{l-1,l}^{(l)}$ and $Y_l = F_{l+1,l}^{(l)}$, and then rewriting (9.125) as

$$F_{0,1}^{(l)} = \Big[A_0 - \Sigma_L^r(\zeta_0)\Big]^{-1} B_0$$

$$F_{1,2}^{(l)} = \Big[A_1 - B_0^\dagger F_{0,1}^{(l)}\Big]^{-1} B_1$$

$$\vdots$$

$$X_l = F_{l-1,l}^{(l)} = \Big[A_{l-1} - B_{l-2}^\dagger F_{l-2,l-1}^{(l)}\Big]^{-1} B_{l-1}, \tag{9.131}$$

we find that (9.131) is equivalent to the continued-fraction equation for X_l in (9.122). On the other hand, (9.127) is transformed into

$$Y_l = F_{l+1,l}^{(l)} = \Big[A_{l+1} - B_{l+1}F_{l+2,l+1}^{(l)}\Big]^{-1} B_l^\dagger$$

$$\vdots$$

$$F_{m,m-1}^{(l)} = \Big[A_m - B_m F_{m+1,m}^{(l)}\Big]^{-1} B_{m-1}^\dagger$$

$$F_{m+1,m}^{(l)} = \Big[A_{m+1} - \Sigma_R^r(\zeta_{m+1})\Big]^{-1} B_m^\dagger, \tag{9.132}$$

which gives the expression of Y_l in (9.122).

Finally, X_0 and Y_{m+1} in the particular cases of $l = 0$ and $m+1$, (9.123), are determined using the equations

$$(A_0 - \Sigma_L^r(\zeta_0))G_T^r(\zeta_0, \zeta_0) - B_0 G_T^r(\zeta_1, \zeta_0) = I$$
$$-B_m^\dagger G_T^r(\zeta_m, \zeta_{m+1}) + (A_{m+1} - \Sigma_R^r(\zeta_{m+1}))G_T^r(\zeta_{m+1}, \zeta_{m+1}) = I \tag{9.133}$$

which come from (9.124) for $l = 0$ and $l = m + 1$, respectively. $\qquad\square$

From (9.121) – (9.123), one sees that $B_{l-1}^{\dagger} X_l$ ($B_l Y_l$) for $l = 1, 2, ..., m$ correspond to the continuation of the self-energy term $\sum_L^r(\zeta_0)$ ($\sum_R^r(\zeta_{m+1})$) defined on the left- (right-)electrode surface [see Eq. (9.68)] to the transition region. It should be noted that these continued self-energy terms $\{X_l\}$ can be sequentially computed such that

$$X_0 = (B_{-1}^{\dagger})^{-1} \sum_L^r(\zeta_0)$$
$$X_1 = (A_0 - B_{-1}^{\dagger} X_0)^{-1} B_0$$
$$X_2 = (A_1 - B_0^{\dagger} X_1)^{-1} B_1$$
$$\vdots$$
$$X_{m+1} = (A_m - B_{m-1}^{\dagger} X_m)^{-1} B_m, \tag{9.134}$$

and similarly, $\{Y_l\}$ are obtainable as an iterative series,

$$Y_{m+1} = (B_{m+1})^{-1} \sum_R^r(\zeta_{m+1})$$
$$Y_m = (A_{m+1} - B_{m+1} Y_{m+1})^{-1} B_m^{\dagger}$$
$$Y_{m-1} = (A_m - B_m Y_m)^{-1} B_{m-1}^{\dagger}$$
$$\vdots$$
$$Y_0 = (A_1 - B_1 Y_1)^{-1} B_0^{\dagger}, \tag{9.135}$$

which allows us to calculate all $G_T^r(\zeta_l, \zeta_l)$'s of (9.121) by a linear scaling operation (order-N calculation procedure) with a limited computational cost. It is also noted that (9.134) and (9.135) are stably computed without involving error accumulation. This can be seen from the fact that these equations resemble the continued-fraction equations for the ratio matrices R^{ref} and R^{tra} [see Eqs. (6.72) and (6.75)], in which errors due to the appearance of evanescent waves are eliminated by introducing the ratios of these waves at two successive grid points.

Once $G_T^r(\zeta_l, \zeta_l)$, X_l and Y_l for any l ($0 \le l \le m + 1$) are known,

$$G_T^r(\zeta_{l-1}, \zeta_l) = X_l G_T^r(\zeta_l, \zeta_l)$$
$$G_T^r(\zeta_{l+1}, \zeta_l) = Y_l G_T^r(\zeta_l, \zeta_l) \tag{9.136}$$

are also known from (9.129), and eventually all of the elements $\{G_T^r(\zeta_k, \zeta_l)\}$ are determined through the relationships between two or three adjacent terms, (9.125) and (9.127).

For later convenience, we give the equality existing among $F_{i,j}^{(k)}$'s defined by (9.130).

$$X_l = F^{(l)}_{l-1,l} = F^{(l+1)}_{l-1,l} = \cdots = F^{(m+1)}_{l-1,l} \quad (1 \leq l \leq m+1)$$
$$Y_l = F^{(l)}_{l+1,l} = F^{(l-1)}_{l+1,l} = \cdots = F^{(0)}_{l+1,l} \quad (0 \leq l \leq m). \tag{9.137}$$

These equations are of some practical use for computing off-diagonal elements of $\{G^r_T(\zeta_k, \zeta_l)\}$. Here we verify, for example, $F^{(l)}_{l-1,l} = F^{(l+1)}_{l-1,l}$. Equations (9.125) with the replacement of l by $l+1$ are rewritten, similarly to (9.131), in terms of $F^{(l+1)}_{i,j}$ as

$$F^{(l+1)}_{0,1} = \left[A_0 - \Sigma^r_L(\zeta_0)\right]^{-1} B_0$$
$$F^{(l+1)}_{1,2} = \left[A_1 - B^\dagger_0 F^{(l+1)}_{0,1}\right]^{-1} B_1$$
$$\vdots$$
$$F^{(l+1)}_{l-1,l} = \left[A_{l-1} - B^\dagger_{l-2} F^{(l+1)}_{l-2,l-1}\right]^{-1} B_{l-1}$$
$$F^{(l+1)}_{l,l+1} = \left[A_l - B^\dagger_{l-1} F^{(l+1)}_{l-1,l}\right]^{-1} B_l, \tag{9.138}$$

and then the upper l equations yield a continued-fraction representation for $F^{(l+1)}_{l-1,l}$ in the form of

$$F^{(l+1)}_{l-1,l} = \left[A_{l-1} - B^\dagger_{l-2}\left[A_{l-2} - \cdots - B^\dagger_0\left[A_0 - \Sigma^r_L(\zeta_0)\right]^{-1} B_0\right.\right.$$
$$\left.\left. \cdots \right]^{-1} B_{l-3}\right]^{-1} B_{l-2}\right]^{-1} B_{l-1}, \tag{9.139}$$

which is equal to the expression of X_l $(= F^{(l)}_{l-1,l})$ in (9.122). [The last equation in (9.138) is irrelevant to the present discussion.] $\qquad\square$

In the rest of this section, we focus on discussing the relationship between the retarded Green's function and scattering wave function. Let us denote the scattering wave function for the j-th incident wave $\Phi^{in}_{L,j}$ $(\Phi^{in}_{R,j})$ incoming from deep inside the left (right) electrode as $\Psi_{L,j}$ $(\Psi_{R,j})$, where the incident waves are considered to include evanescent waves as well as ordinary propagating waves; more precisely, $\{\Phi^{in}_{L,j}\}$ $(\{\Phi^{in}_{R,j}\})$ is here taken to be a set of the N generalized Bloch states consisting of rightward (leftward) propagating Bloch waves and decaying evanescent waves toward the right (left) side, which are solutions of the $2N$-dimensional generalized eigenvalue equation [see Eq. (7.1)]. We define the N-dimensional matrix which gathers the corresponding scattering wave functions at ζ_l by $U_L(\zeta_l)$ $(U_R(\zeta_l))$, i.e.,

$$U_A(\zeta_l) = \left[\Psi_{A,1}(\zeta_l), \Psi_{A,2}(\zeta_l), ..., \Psi_{A,N}(\zeta_l)\right] \quad (A = L \text{ and } R). \quad (9.140)$$

We further define the ratio matrices for U_L and U_R by

$$S_L(\zeta_l) = U_L(\zeta_{l+1})U_L(\zeta_l)^{-1}$$
$$S_R(\zeta_l) = U_R(\zeta_{l-1})U_R(\zeta_l)^{-1}. \quad (9.141)$$

Here and hereafter, we suppress the dependence of $\Psi_{A,j}$, U_A and S_A $(A = L$ and $R)$ on E and $\mathbf{k}_{\parallel}$ for simplicity.

It can be shown that for $0 \leq l \leq m+1$,

$$X_l = S_R(\zeta_l) \quad \text{and} \quad Y_l = S_L(\zeta_l). \quad (9.142)$$

The proof is as follows. The case of incident waves incoming from the left electrode is considered. $\{U_L(\zeta_k)\}$ obeys the Kohn–Sham equation

$$-B_k^\dagger U_L(\zeta_k) + A_{k+1}U_L(\zeta_{k+1}) - B_{k+1}U_L(\zeta_{k+2}) = 0 \ (-\infty < k < \infty). \quad (9.143)$$

The multiplication of this equation by $U_L(\zeta_{k+1})^{-1}$ from the right side yields

$$S_L(\zeta_k) = \left[A_{k+1} - B_{k+1}S_L(\zeta_{k+1})\right]^{-1}B_k^\dagger \quad (9.144)$$

and the successive use of (9.144) from $k = l$ to m leads to

$$S_L(\zeta_l) = \left[A_{l+1} - B_{l+1}\left[A_{l+2} - \cdots - B_m\left[A_{m+1} - B_{m+1}S_L(\zeta_{m+1})\right]^{-1}B_m^\dagger\right.\right.$$
$$\left.\left.\cdots\right]^{-1}B_{l+2}^\dagger\right]^{-1}B_{l+1}^\dagger\right]^{-1}B_l^\dagger. \quad (9.145)$$

Recall that the scattering boundary condition inside the right electrode, (9.110), reads as

$$U_L(\zeta_l) = Q^{tra}(\zeta_l)T \quad (l \geq m+1), \quad (9.146)$$

in which T is the transmission-coefficient matrix (9.114). Then the relation

$$S_L(\zeta_l) = R^{tra}(\zeta_{l+1}) \quad (l \geq m+1) \quad (9.147)$$

is easily found to hold by the definition of the ratio matrix of $R^{tra}(\zeta_{l+1})$ [see Eq. (7.2)], i.e.,

$$R^{tra}(\zeta_{l+1}) = Q^{tra}(\zeta_{l+1})Q^{tra}(\zeta_l)^{-1} = U_L(\zeta_{l+1})U_L(\zeta_l)^{-1}. \quad (9.148)$$

Thus, from (9.70) and (9.147), one sees that $S_L(\zeta_l)$ of (9.145) is identical to Y_l in (9.122). A similar argument in the case of incident waves from the right electrode derives the other equation, $X_l = S_R(\zeta_l)$. $\qquad\square$

When (9.141) and (9.142) are combined with (9.104) [or (9.105)] and Theorem 9.4, there emerges a calculation procedure for scattering wave functions, which is equivalent to the recursion-transfer-matrix method [Hirose and Tsukada (1994, 1995); Hirose *et al.* (2004)]. This is stated as follows. For example, in the case of an incident wave Φ_L^{in} incoming from the left electrode, the value of the scattering wave function at ζ_0, $\Psi_L(\zeta_0)$, is calculated by (9.104) and Theorem 9.4, namely,

$$\Psi_L(\zeta_0) = iG_T^r(\zeta_0, \zeta_0)\Gamma_L(\zeta_0)\Phi_L^{in}(\zeta_0) \tag{9.149}$$

with

$$
\begin{aligned}
G_T^r(\zeta_0, \zeta_0) &= \left[A_0 - B_{-1}^\dagger X_0 - B_0 Y_0 \right]^{-1} \\
X_0 &= (B_{-1}^\dagger)^\dagger \Sigma_L^r(\zeta_0) \\
Y_0 &= \left[A_1 - B_1 \left[A_2 - \cdots - B_m \left[A_{m+1} - \Sigma_R^r(\zeta_{m+1}) \right]^{-1} B_m^\dagger \right. \right. \\
&\qquad \left. \left. \cdots \right]^{-1} B_2^\dagger \right]^{-1} B_1^\dagger \right]^{-1} B_0^\dagger,
\end{aligned}
\tag{9.150}
$$

and subsequently, using (9.141) and (9.142), the values in the transition region are iteratively computed as

$$
\Psi_L(\zeta_1) = Y_0 \Psi_L(\zeta_0), \quad \Psi_L(\zeta_2) = Y_1 \Psi_L(\zeta_1),
$$
$$
\cdots, \Psi_L(\zeta_{m+1}) = Y_m \Psi_L(\zeta_m). \tag{9.151}
$$

Here, $\Sigma_{\{L,R\}}^r$, Γ_L, and Y_l ($l = m, m-1, ..., 0$) are obtained from (9.70), (9.98), and (9.135), respectively. The present Green's function formalism incorporates *crystalline* electrodes and norm-conserving nonlocal pseudopotentials. This means that the recursion-transfer-matrix method is so extended as to include these proper ingredients of the first-principles calculation procedure in it.

It is now possible to generalize the relations between the retarded Green's functions and scattering wave functions, (9.104) and (9.105), as given in the following theorem.

Theorem 9.5

$$
\begin{aligned}
\Psi_L(\zeta_l) &= iG_T^r(\zeta_l, \zeta_0)\Gamma_L(\zeta_0)\Phi_L^{in}(\zeta_0) \\
\Psi_R(\zeta_l) &= iG_T^r(\zeta_l, \zeta_{m+1})\Gamma_R(\zeta_{m+1})\Phi_R^{in}(\zeta_{m+1}) \\
&\qquad (0 \leq l \leq m+1)
\end{aligned}
\tag{9.152}
$$

Proof. By the definition of S_L, (9.141), and the use of (9.130), (9.137) and (9.142),

$$
\begin{aligned}
\Psi_L(\zeta_l) &= S_L(\zeta_{l-1})\Psi_L(\zeta_{l-1}) \\
&= S_L(\zeta_{l-1})S_L(\zeta_{l-2})\cdots S_L(\zeta_0)\Psi_L(\zeta_0) \\
&= Y_{l-1}Y_{l-2}\cdots Y_0\Psi_L(\zeta_0) \\
&= F^{(0)}_{l,l-1}F^{(0)}_{l-1,l-2}\cdots F^{(0)}_{1,0}\Psi_L(\zeta_0) \\
&= G^r_T(\zeta_l,\zeta_0)G^r_T(\zeta_{l-1},\zeta_0)^{-1}G^r_T(\zeta_{l-1},\zeta_0)G^r_T(\zeta_{l-2},\zeta_0)^{-1} \\
&\quad \times \cdots \times G^r_T(\zeta_1,\zeta_0)G^r_T(\zeta_0,\zeta_0)^{-1}\Psi_L(\zeta_0) \\
&= G^r_T(\zeta_l,\zeta_0)G^r_T(\zeta_0,\zeta_0)^{-1}\Psi_L(\zeta_0) \\
&= iG^r_T(\zeta_l,\zeta_0)\Gamma_L(\zeta_0)\Phi^{in}_L(\zeta_0)
\end{aligned}
\tag{9.153}
$$

is derived. The last step follows from (9.104). The expression of $\Psi_R(\zeta_l)$ in (9.152) is similarly derived. $\qquad\square$

Finally, we examine the representation of the retarded Green's function by scattering wave functions. This representation is useful in transport calculations relying on the Lippmann–Schwinger equation, which will be discussed in the next chapter. In the one-dimensional scattering theory treating a second-order differential equation of the form

$$
\left[E - H(z,\frac{d}{dz})\right]\psi(z) = 0
\tag{9.154}
$$

with

$$
H(z,\frac{d}{dz}) = -\frac{1}{2}\frac{d^2}{dz^2} + v(z),
\tag{9.155}
$$

it is well known that the retarded Green's function of this system, $G^r(z,z';E)$, can be made up of two solutions of the scattering waves, $\psi_L(z)$ and $\psi_R(z)$, which are the solutions of $[E-H]\psi = 0$ corresponding to the incident waves incoming from the left and right sides, respectively, as follows [see, for instance, Lang (1995)].

$$
G^r(z,z';E) = \begin{cases} \frac{1}{w}\psi_R(z)\psi_L(z') & \text{for } z \leq z' \\[2mm] \frac{1}{w}\psi_L(z)\psi_R(z') & \text{for } z \geq z' , \end{cases}
\tag{9.156}
$$

where w is a constant corresponding to half of the Wronskian

$$
w = \frac{1}{2}\begin{vmatrix} \psi_R(z) & \psi_L(z) \\ \frac{d}{dz}\psi_R(z) & \frac{d}{dz}\psi_L(z) \end{vmatrix}.
\tag{9.157}
$$

This procedure for obtaining the retarded Green's function in the one-dimensional *continuous* space can be simply extended to the case of the one-dimensional *discretized* space when the central finite-difference formula is adopted for the operator $-\frac{1}{2}\frac{d^2}{dz^2}$. That is, we still have

$$G^r(z_k, z_l; E) = \begin{cases} \frac{1}{w}\psi_R(z_k)\psi_L(z_l) & \text{for } z_k \leq z_l \\ \frac{1}{w}\psi_L(z_k)\psi_R(z_l) & \text{for } z_k \geq z_l \end{cases} \tag{9.158}$$

with

$$w = \frac{1}{2h_z}\begin{vmatrix} \psi_R(z_l) & \psi_L(z_l) \\ \frac{1}{h_z}(\psi_R(z_{l+1}) - \psi_R(z_l)) & \frac{1}{h_z}(\psi_L(z_{l+1}) - \psi_L(z_l)) \end{vmatrix}$$
$$= b\left[\psi_R(z_{l+1})\psi_L(z_l) - \psi_L(z_{l+1})\psi_R(z_l)\right]. \tag{9.159}$$

Equation (9.158) is derived by observing that the l-th columnar vector of $\{G^r(z_k, z_l; E)\}$, $k = ..., l-1, l, l+1, ...$, given by (9.158) satisfies

$$\frac{1}{w}\begin{bmatrix} \ddots & \ddots & \ddots & & 0 \\ & -b & a_{l-2} & -b \\ & & -b & a_{l-1} & -b \\ & & & -b & a_l & -b \\ & & & & -b & a_{l+1} & -b \\ & & & & & -b & a_{l+2} & -b \\ 0 & & & & & \ddots & \ddots & \ddots \end{bmatrix}\begin{bmatrix} \vdots \\ \psi_R(z_{l-2})\psi_L(z_l) \\ \psi_R(z_{l-1})\psi_L(z_l) \\ \psi_R(z_l)\psi_L(z_l) \\ \psi_R(z_l)\psi_L(z_{l+1}) \\ \psi_R(z_l)\psi_L(z_{l+2}) \\ \vdots \end{bmatrix} = \begin{bmatrix} \vdots \\ 0 \\ 0 \\ 1 \\ 0 \\ 0 \\ \vdots \end{bmatrix}, \tag{9.160}$$

since $\psi_{\{R,L\}}$ obey the discretized equation of (9.154), namely,

$$-b\psi_{\{R,L\}}(z_{l-1}) + a_l\psi_{\{R,L\}}(z_l) - b\psi_{\{R,L\}}(z_{l+1}) = 0, \tag{9.161}$$

where,

$$a_l = E - \frac{1}{h_z^2} - v(z_l) \quad \text{and} \quad b = -\frac{1}{2h_z^2}, \tag{9.162}$$

and h_z is the grid spacing. In addition, we can show that w of (9.159) is l-independent using (9.161).

Is it possible to construct the retarded Green's function in terms of scattering wave functions in a more general case of a three-dimensional system analogously to (9.156) and (9.158)? The following theorem gives an answer. Here, we use again a generalized z-coordinate ζ_l instead of z_l (see Fig. 9.1), and treat the retarded Green's function of the whole system portioned to the transition region, $G_T^r(\zeta_k, \zeta_l; E, \mathbf{k}_{\parallel})$.

Theorem 9.6 *For $0 \leq k,\ l \leq m+1$,*

$$G_T^r(\zeta_k, \zeta_l; E, \mathbf{k}_{\parallel}) = \begin{cases} U_R(\zeta_k)U_R(\zeta_l)^{-1}\Lambda(\zeta_l)^{-1} & \text{for}\ \ \zeta_k \leq \zeta_l \\ U_L(\zeta_k)U_L(\zeta_l)^{-1}\Lambda(\zeta_l)^{-1} & \text{for}\ \ \zeta_k \geq \zeta_l \end{cases} \tag{9.163}$$

with

$$\Lambda(\zeta_l) = A_l - B_{l-1}^{\dagger}U_R(\zeta_{l-1})U_R(\zeta_l)^{-1} - B_l U_L(\zeta_{l+1})U_L(\zeta_l)^{-1}, \tag{9.164}$$

where $U_{\{R,L\}}$ are the matrices consisting of the wave functions $\Psi_{\{R,L\},j}$'s, defined by (9.140).

Proof. The retarded Green's function is constructed by the outwardly propagating and decreasing waves. Taking it into account, for the l-th column of $\hat{G}_T^r(E, \mathbf{k}_{\parallel})$, i.e., $[G_T^r(\zeta_0, \zeta_l), G_T^r(\zeta_1, \zeta_l), ..., G_T^r(\zeta_{m+1}, \zeta_l)]^t$, we assume

$$G_T^r(\zeta_k, \zeta_l) = \begin{cases} U_R(\zeta_k)\Delta_l & \text{for}\ \ 0 \leq k \leq l-1 \\ D_l & \text{for}\ \ k = l \\ U_L(\zeta_k)\Delta_l' & \text{for}\ \ l+1 \leq k \leq m+1, \end{cases} \tag{9.165}$$

where Δ_l, D_l and Δ_l' are unknown block matrices which will be so determined that $\{G_T^r(\zeta_k, \zeta_l)\}$ will satisfy (9.124), or equivalently (9.125) – (9.127). Since $\{U_{\{R,L\}}(\zeta_k)\}$ obeys the discretized Kohn–Sham equation

$$-B_{k-1}^{\dagger}U_{\{R,L\}}(\zeta_{k-1}) + A_k U_{\{R,L\}}(\zeta_k) - B_k U_{\{R,L\}}(\zeta_{k+1}) = 0, \tag{9.166}$$

$\{G_T^r(\zeta_k, \zeta_l)\}$ given by (9.163) is found to evidently satisfy all the equations of (9.125) – (9.127), except for three equations: the bottom line of (9.125), (9.126) and the top line of (9.127). [It may be necessary to comment that the top (bottom) line of (9.125) ((9.127)) is satisfied by $\{U_R(\zeta_0), U_R(\zeta_1)\}$ ($\{U_L(\zeta_m), U_L(\zeta_{m+1})\}$); in fact this postulation is verified by (9.122), (9.123), (9.141) and (9.142).] The three equations in question now read as

$$-B_{l-2}^{\dagger}U_R(\zeta_{l-2})\Delta_l + A_{l-1}U_R(\zeta_{l-1})\Delta_l - B_{l-1}D_l = 0$$

$$-B_{l-1}^{\dagger}U_R(\zeta_{l-1})\Delta_l + A_l D_l - B_l U_L(\zeta_{l+1})\Delta_l' = I$$

$$-B_l^{\dagger}D_l + A_{l+1}U_L(\zeta_{l+1})\Delta_l' - B_{l+1}U_L(\zeta_{l+2})\Delta_l' = 0. \tag{9.167}$$

Thus, using (9.166), we find that (9.167) is satisfied by

$$\Delta_l = U_R(\zeta_l)^{-1}D_l, \quad \Delta_l' = U_L(\zeta_l)^{-1}D_l, \tag{9.168}$$

where $D_l = \Lambda(\zeta_l)^{-1}$ with $\Lambda(\zeta_l)$ given by (9.164). □

We note that the diagonal element $G_T^r(\zeta_l, \zeta_l; E, \mathbf{k}_\parallel)$ of (9.163) is $\Lambda(\zeta_l)^{-1}$; indeed, the inverse of (9.164) is equivalent to (9.121) of Theorem 9.4 by virtue of (9.141) and (9.142). This means that Theorem 9.6 includes Theorem 9.4. We further note that if $U_R(\zeta_l)$ and $U_L(\zeta_l)$ were commutable, (9.163) could be rewritten in a similar form to (9.158) as

$$G_T^r(\zeta_k, \zeta_l; E, \mathbf{k}_\parallel) = \begin{cases} U_R(\zeta_k)U_L(\zeta_l)W(\zeta_l)^{-1} & \text{for} \quad \zeta_k \leq \zeta_l \\ \\ U_L(\zeta_k)U_R(\zeta_l)W(\zeta_l)^{-1} & \text{for} \quad \zeta_k \geq \zeta_l \end{cases} \tag{9.169}$$

with

$$\begin{aligned} W(\zeta_l) &= \Lambda(\zeta_l)U_R(\zeta_l)U_L(\zeta_l) \\ &= B_l[U_R(\zeta_{l+1})U_L(\zeta_l) - U_L(\zeta_{l+1})U_R(\zeta_l)]. \end{aligned} \tag{9.170}$$

In addition, it should be remarked that a feature of the retarded Green's function of (9.163) is 'separability' with respect to the variables ζ_k and ζ_l; this seems to be very attractive when employing (9.163) as the Green's function of an unperturbed system in the approach based on the Lippmann–Schwinger equation, in combination with iterative algorithms for solving simultaneous linear equations (see the next chapter).

Chapter 10

Calculation Method Based on the Lippmann–Schwinger Equation

Shigeru Tsukamoto and Kikuji Hirose

10.1 The Lippmann–Schwinger Equation

In previous chapters, we have established that the overbridging boundary-matching (OBM) method is useful in describing electron-transport properties of the nanostructures sandwiched between a pair of electrodes. In this chapter, we address another self-consistent calculation method for computing scattering wave functions, i.e., the method using the Lippmann–Schwinger equation. The methodology based on the Lippmann–Schwinger equation has been proposed by Lang *et al.* For contributions in this direction, see, for example, Lang and Williams (1978), Lang (1995), Lang and Avouris (1998, 2000, 2001), Di Ventra *et al.* (2000, 2001, 2002), Di Ventra and Lang (2001), Kobayashi *et al.* (2001), Tsukamoto and Hirose (2002), Tsukamoto *et al.* (2002) and Chen and Di Ventra (2003).

We are interested in finding eigenstates (scattering wave functions) of the Hamiltonian of an open system of 'left electrode–transition region–right electrode' (see Fig. 6.2), under the restriction of scattering boundary conditions inside the left and right electrodes. In the formulation, the electrodes are assumed to be semi-infinite *crystalline* bulks, in general. For a definitive view, we consider the case of a scattering wave function $\psi(\boldsymbol{r})$ for an energy E whose incident wave comes from the left electrode, i.e., a wave function $\psi(\boldsymbol{r})$ that obeys the Kohn–Sham equation,

$$H\psi(\boldsymbol{r}) = E\psi(\boldsymbol{r}), \qquad (10.1)$$

with the Hamiltonian (a local effective potential is assumed for the time being),

$$H = -\frac{1}{2}\nabla^2 + v_{eff}(\boldsymbol{r}) \tag{10.2}$$

and satisfies the boundary conditions [see Eqs. (6.14) and (6.15)],

$$\psi(\boldsymbol{r}_{||}, z) = \begin{cases} \phi^{in}(\boldsymbol{r}_{||}, z) + \sum_{i=1}^{N} r_i \phi_i^{ref}(\boldsymbol{r}_{||}, z) & \text{in the left electrode} \\ & (z \leq z_0) \\ \sum_{i=1}^{N} t_i \phi_i^{tra}(\boldsymbol{r}_{||}, z) & \text{in the right electrode} \\ & (z \geq z_{m+1}) . \end{cases} \tag{10.3}$$

Here, $\boldsymbol{r} = (\boldsymbol{r}_{||}, z)$, and ϕ^A ($A = in$, ref and tra) are generalized Bloch states peculiar to the electrodes (see Fig. 6.2 and Section 6.3). We implicitly assume that energy E belongs to the continuous spectrum of H.

Let us decompose H into an unperturbed part, H^0, and a perturbation δv as

$$H = H^0 + \delta v, \tag{10.4}$$

where

$$H^0 = -\frac{1}{2}\nabla^2 + v_{eff}^0(\boldsymbol{r}) \tag{10.5}$$

is such that one can easily solve its eigenvalue equation,

$$H^0 \psi^0(\boldsymbol{r}) = E^0 \psi^0(\boldsymbol{r}), \tag{10.6}$$

for E^0 equal to E, imposing the same boundary conditions on an unperturbed wave function ψ^0 as (10.3), i.e.,

$$\psi^0(\boldsymbol{r}_{||}, z) = \begin{cases} \phi^{in}(\boldsymbol{r}_{||}, z) + \sum_{i=1}^{N} r_i^0 \phi_i^{ref}(\boldsymbol{r}_{||}, z) & \text{in the left electrode} \\ & (z \leq z_0) \\ \sum_{i=1}^{N} t_i^0 \phi_i^{tra}(\boldsymbol{r}_{||}, z) & \text{in the right electrode} \\ & (z \geq z_{m+1}) . \end{cases} \tag{10.7}$$

Here, the superscript '0' accounts for quantities of the unperturbed system. For this purpose, we assume the unperturbed system to be equipped with the same electrodes as those in the perturbed system described by H, and therefore, the perturbation δv ($= H - H^0$) to be nonzero only in the transition region ($z_0 < z < z_{m+1}$). In terms of H^0 and δv, (10.1) is rewritten as

$$(E - H^0)\psi(\boldsymbol{r}) = \delta v(\boldsymbol{r})\psi(\boldsymbol{r}), \tag{10.8}$$

and once the retarded Green's function $G_T^{r0}(\boldsymbol{r}, \boldsymbol{r}'; E)$ in the transition region associated with the unperturbed part H^0 is known, (10.8) is put into the Lippmann–Schwinger form,

$$\psi(\boldsymbol{r}) = \psi^0(\boldsymbol{r}) + \int_{z_0}^{z_{m+1}} dz' \int dr'_\| G_T^{r0}(\boldsymbol{r}, \boldsymbol{r}'; E)\delta v(\boldsymbol{r}')\psi(\boldsymbol{r}'). \qquad (10.9)$$

Equation (10.9) is an integral equation of the second kind of the Fredholm type with respect to the unknown function $\psi(\boldsymbol{r})$. Using (10.6) with $E^0 = E$ and the definition of the Green's function, i.e.,

$$(E - H^0)G_T^{r0}(\boldsymbol{r}, \boldsymbol{r}'; E) = \delta(\boldsymbol{r}, \boldsymbol{r}'), \qquad (10.10)$$

it is easy to show that the solution of (10.9) obeys (10.8), and hence (10.1), since multiplying $(E - H^0)$ from the left side of (10.9) obviously yields (10.8). Furthermore, one sees that the solution of (10.9) satisfies the scattering boundary conditions (10.3), recalling that the retarded Green's function $G_T^{r0}(\boldsymbol{r}, \boldsymbol{r}'; E)$ is constructed by the outwardly propagating and decreasing waves, $\{\phi_i^{ref}(\boldsymbol{r}_\|, z)\}$ and $\{\phi_i^{tra}(\boldsymbol{r}_\|, z)\}$, in the electrode regions. This point is obvious in the one-dimensional example of (9.156) as follows. In this case, put that $N = 1$ in (10.3) and (10.7). The retarded Green's function (9.156), regarded as that associated with the unperturbed H^0, now reads as

$$G_T^{r0}(z, z'; E) = \begin{cases} \dfrac{1}{w}\psi_R^0(z)\psi_L^0(z') & \text{for } z \le z' \\[2ex] \dfrac{1}{w}\psi_L^0(z)\psi_R^0(z') & \text{for } z \ge z', \end{cases} \qquad (10.11)$$

with

$$\begin{aligned} \psi_R^0(z) &= \phi_1^{ref}(z) \quad \text{for } z \le z_0, \\ \psi_L^0(z) &= \phi_1^{tra}(z) \quad \text{for } z \ge z_{m+1}. \end{aligned} \qquad (10.12)$$

Thus, the Lippmann–Schwinger equation (10.9) is found to give the proper boundary conditions such that

$$\psi(z) = \begin{cases} \phi^{in}(z) + r_1\phi_1^{ref}(z) & \text{for } z \le z_0, \\ t_1\phi_1^{tra}(z) & \text{for } z \ge z_{m+1}, \end{cases} \qquad (10.13)$$

where

$$r_1 = r_1^0 + \frac{1}{w} \int_{z_0}^{z_{m+1}} dz' \phi_1^{tra}(z') \delta v(z') \psi(z')$$

$$t_1 = t_1^0 + \frac{1}{w} \int_{z_0}^{z_{m+1}} dz' \phi_1^{ref}(z') \delta v(z') \psi(z').$$

(10.14)

Consequently, we see that an integral equation of the Lippmann–Schwinger form (10.9) provides a unified description of the Kohn–Sham equation (10.1) and the boundary conditions (10.3).

It is noted that one can adopt any unperturbed system, as long as the solution $\psi^0(r)$ of its eigenvalue equation (10.6) satisfies the boundary conditions (10.7). Often, the system consisting of only bare left and right electrodes (i.e., with the empty transition region) is chosen as an unperturbed one. This way, by using the same unperturbed Green's function of the bare electrode system, G_T^{r0}, which is calculated *once*, scattering wave functions $\psi(r)$ can be efficiently evaluated from (10.9) for a variety of structures of nanoscale junctions set up in the transition region. In addition, for a similar reason, the Lippmann–Schwinger equation (10.9) is utilized in the implementation of self-consistent calculations for the convergence of electronic states in an infinitely open system [Tsukamoto and Hirose (2002)].

The extension of the present scheme to the case of including norm-conserving nonlocal pseudopotentials v_{NL} follows exactly the same procedures. In this case, the Kohn–Sham equation is written as

$$-\frac{1}{2}\nabla^2 \psi(r) + v_{eff}(r)\psi(r) + \int dr' v_{NL}(r,r')\psi(r') = E\psi(r), \qquad (10.15)$$

and it is rewritten in terms of the Lippmann–Schwinger equation as

$$\psi(r) = \psi^0(r) + \int dr' dr'' G_T^{r0}(r,r';E)\delta v(r',r'')\psi(r''), \qquad (10.16)$$

where

$$\delta v(r,r') = \left[v_{eff}(r) - v_{eff}^0(r)\right]\delta(r-r') + v_{NL}(r,r') - v_{NL}^0(r,r'). \qquad (10.17)$$

The perturbed wave function $\psi(r)$ and unperturbed one $\psi^0(r)$ are imposed by the same boundary conditions inside the left and right electrodes [(10.3) and (10.7)], and the perturbation $\delta v(r,r')$ is then zero in the electrode regions, i.e., $\delta v(r,r') \neq 0$ only for $z_0 < z, z' < z_{m+1}$.

We move on to the Lippmann–Schwinger formalism within the framework of the real-space finite-difference approach. The discretization of the space into grid points r_k makes no alterations to the essence of the above-mentioned scheme. Therefore we only collect the necessary formulas in a discrete real-space matrix representation. A generalized z-coordinate ζ_k is used instead of z_k (see Section 7.1); thus, the formalism incorporates norm-conserving nonlocal pseudopotentials in it. In practice, for saving computational costs, the nonlocal pseudopotentials are employed in a separable form of Kleinman and Bylander [see Eq. (1.23)].

The Kohn–Sham equation for the perturbed system [cf. Eq. (6.7)] is

$$-B(\zeta_{k-1})^{\dagger}\Psi(\zeta_{k-1}) + [E - H(\zeta_k)]\,\Psi(\zeta_k) - B(\zeta_k)\Psi(\zeta_{k+1}) = 0$$
$$(k = -\infty, \dots, -1, 0, 1, \dots, \infty). \quad (10.18)$$

Here, we assumed that the Hamiltonian of the whole system is represented in a block-tridiagonal form, for simplicity. The scattering boundary conditions in the case of an incident wave $\Phi^{in}(\zeta_k)$ from the left electrode [cf. Eqs. (6.14) and (6.15)] are

$$\Psi(\zeta_k) = \begin{cases} \Psi^{in}(\zeta_k) + \displaystyle\sum_{i=1}^{N} r_i \Phi_i^{ref}(\zeta_k) & \text{in the left electrode } (k \leq 0) \\ \displaystyle\sum_{i=1}^{N} t_i \Phi_i^{tra}(\zeta_k) & \text{in the right electrode } (k \geq m+1). \end{cases}$$
$$(10.19)$$

The Lippmann–Schwinger equation equivalent to the unification of (10.18) and (10.19) is

$$\Psi(\zeta_k) = \Psi^0(\zeta_k) + \sum_{l,l'=1}^{m} G_T^{r0}(\zeta_k, \zeta_l)\delta V(\zeta_l, \zeta_{l'})\Psi(\zeta_{l'}) \quad (k = 0, 1, \dots, m+1).$$
$$(10.20)$$

The Kohn–Sham equation and the relevant boundary conditions for the unperturbed system are

$$-B^0(\zeta_{k-1})^{\dagger}\Psi^0(\zeta_{k-1}) + [E - H^0(\zeta_k)]\,\Psi^0(\zeta_k) - B^0(\zeta_k)\Psi^0(\zeta_{k+1}) = 0$$
$$(k = -\infty, \dots, -1, 0, 1, \dots, \infty) \quad (10.21)$$

with $H^0(\zeta_k) = H(\zeta_k)$ and $B^0(\zeta_k) = B(\zeta_k)$ for $k \leq 0$ and $k \geq m+1$, and

$$\Psi^0(\zeta_k) = \begin{cases} \Phi^{in}(\zeta_k) + \displaystyle\sum_{i=1}^{N} r_i^0 \Phi_i^{ref}(\zeta_k) & \text{in the left electrode } (k \leq 0) \\ \displaystyle\sum_{i=1}^{N} t_i^0 \Phi_i^{tra}(\zeta_k) & \text{in the right electrode } (k \geq m+1). \end{cases}$$

(10.22)

The perturbation (the difference between the Hamiltonian matrices of the perturbed and unperturbed systems), $\delta V(\zeta_l, \zeta_{l'})$, is as follows: only for the transition region $(1 \leq l, l' \leq m)$, $\delta V(\zeta_l, \zeta_{l'}) \neq 0$, and in this region,

$$\delta V(\zeta_l, \zeta_{l'}) = \begin{cases} H(\zeta_l) - H^0(\zeta_l) & \text{for } l' = l \\ B(\zeta_l) - B^0(\zeta_l) & \text{for } l' = l+1 \\ B(\zeta_{l-1})^\dagger - B^0(\zeta_{l-1})^\dagger & \text{for } l' = l-1 \\ 0 & \text{otherwise .} \end{cases}$$

(10.23)

Here, for simplicity we have assumed the Hamiltonian matrices of both the perturbed and unperturbed systems to be block-tridiagonal, in which all of the block matrices, $H^{(0)}(\zeta_k)$, $B^{(0)}(\zeta_k)$, $\delta V(\zeta_l, \zeta_{l'})$, and $G_T^{r0}(\zeta_k, \zeta_l)$, are N dimensional, as are the vectors $\Psi^{(0)}(\zeta_k)$ and $\Phi^A(\zeta_k)$ ($A = in, ref,$ and tra). Furthermore, we have omitted the dependence of these quantities on the energy E and/or the lateral Bloch wave number $\boldsymbol{k}_{\|}$ for the sake of clarity.

To obtain the retarded Green's function of the unperturbed system, $G_T^{r0}(\zeta_k, \zeta_l)$, we can make use of Theorem 9.6 developed in the preceding chapter. Hereafter, expressions (9.163) and (9.164) in Theorem 9.6 are regarded to be those of the unperturbed system, and therefore, all of the quantities are marked by the superscript '0'. It should be emphasized that the retarded Green's function matrix (9.163) is expressed in a *separable* form with respect to the unperturbed quantities, $U_{\{R,L\}}^0(\zeta_k)$ and $U_{\{R,L\}}^0(\zeta_l)^{-1}\Lambda^0(\zeta_l)^{-1}$. This is highly advantageous for iteratively solving the Lippmann–Schwinger equation (10.20) as simultaneous linear equations for unknown vectors $\{\Psi(\zeta_k)\}$ by conjugate-gradient (CG) methods, where the multiplication of matrix and vector is successively required. Note that if the $(m+2)N$-dimensional Green's function G_T^{r0} were non-separable, the amount of each multiplication would be $[(m+2)N]^2$; actually, it is reduced to the order of $(m+2)N^2$ owing to the separability of G_T^{r0}. However, the matrix $G_T^{r0}(\zeta_k, \zeta_l)\delta V(\zeta_l, \zeta_{l'})$ in (10.20) is non-Hermitian, for which the standard CG algorithm is known to slowly converge, and therefore, we have to try out some modified CG algorithms with rapid convergence [Van der Vorst (1992); Sleijpen and Fokkema (1993); Gutknecht (1993)].

For the sake of convenience in numerical computations, we give the following formula concerning the multiplication in (10.20) such as $\sum G_T^{r0}(\zeta_k, \zeta_l) \delta V(\zeta_l, \zeta_{l'}) \Psi(\zeta_{l'})$, which shows that this multiplication is so neatly expressed in terms of the factors X_k^0 and Y_k^0 defined in (9.122) as to realize a linear scaling calculation (X_k and Y_k are now marked by the superscript '0', since they are quantities of the unperturbed system). For $k = 0, 1, \ldots, m+1$,

$$\sum_{l,l'=1}^{m} G_T^{r0}(\zeta_k, \zeta_l) \delta V(\zeta_l, \zeta_{l'}) \Psi(\zeta_{l'}) = \Psi'(\zeta_k) + P_L(\zeta_k) + P_R(\zeta_k), \qquad (10.24)$$

where

$$\Psi'(\zeta_k) = \Lambda^0(\zeta_k)^{-1} \sum_{l=1}^{m} \delta V(\zeta_k, \zeta_l) \Psi(\zeta_l) \qquad (k = 0, 1, \ldots, m+1)$$

$$P_L(\zeta_0) = P_L(\zeta_1) = 0$$

$$P_L(\zeta_k) = U_L^0(\zeta_k) \sum_{l=1}^{k-1} U_L^0(\zeta_l)^{-1} \Psi'(\zeta_l)$$

$$\qquad = Y_{k-1}^0 [P_L(\zeta_{k-1}) + \Psi'(\zeta_{k-1})] \qquad (k = 2, 3, \ldots, m+1)$$

$$P_R(\zeta_{m+1}) = P_R(\zeta_m) = 0$$

$$P_R(\zeta_k) = U_R^0(\zeta_k) \sum_{l=k+1}^{m} U_R^0(\zeta_l)^{-1} \Psi'(\zeta_l)$$

$$\qquad = X_{k+1}^0 [P_R(\zeta_{k+1}) + \Psi'(\zeta_{k+1})] \qquad (k = m-1, \ldots, 1, 0).$$
$$(10.25)$$

Equations (10.24) and (10.25) are easily derived from (9.163) using (9.134) and (9.135). We note again that $\Lambda^0(\zeta_k)^{-1}$ is equal to the diagonal block-matrix element of the retarded Green's function of (9.121), i.e.,

$$\Lambda^0(\zeta_k)^{-1} = G_T^{r0}(\zeta_k, \zeta_k)$$
$$= \left[E - H^0(\zeta_k) - B^0(\zeta_{k-1})^\dagger X_k^0 - B^0(\zeta_k) Y_k^0 \right]^{-1}, \qquad (10.26)$$

and that from the recursive equations (9.134) and (9.135), $\{X_k^0\}$ and $\{Y_k^0\}$ are obtainable as

$$X_0^0 = B^0(\zeta_{-1})^{\dagger^{-1}} \Sigma_L^r(\zeta_0)$$

$$X_k^0 = \left[E - H^0(\zeta_{k-1}) - B^0(\zeta_{k-2})^\dagger X_{k-1}^0 \right]^{-1} B^0(\zeta_{k-1})$$
$$(k = 1, 2, \ldots, m+1) \qquad (10.27)$$

and

$$Y^0_{m+1} = B^0(\zeta_{m+1})^{-1} \textstyle\sum^r_R(\zeta_{m+1})$$
$$Y^0_k = \left[E - H^0(\zeta_{k+1}) - B^0(\zeta_{k+1})Y^0_{k+1}\right]^{-1} B^0(\zeta_m)^\dagger$$
$$(k = m, \dots, 1, 0). \qquad (10.28)$$

So far, we limited our discussion to the case of a perturbed system whose Hamiltonian matrix is block-tridiagonal. However, it is straightforward to extend the present procedure to a more general case of a perturbed system with the Hamiltonian being of a non-block-tridiagonal matrix in the transition region. In fact, (10.24) – (10.28) undergo no change in treating such a system, provided that the block-tridiagonality remains postulated for an unperturbed system. In addition, it should be remarked that (10.24) – (10.28) exhibit a potential power in electron-transport calculations when approximating crystalline electrodes by jellium ones and adopting invented bare electrodes without atoms for an unperturbed system, since $\Lambda^0(\zeta_k)^{-1}$, X^0_k, and Y^0_k in this case can be analytically expressed through a derivation similar to that of the free-electron Green's function demonstrated in Section 9.2.

10.2   Correction of Effective Potentials inside Electrodes at a Finite Bias Voltage

Scanning-tunneling-spectroscopy measurements for metal and semiconductor surfaces usually exhibit nonlinear current–voltage (I–V) characteristics, which closely relate to the local density of states (LDOS) of the materials. It is also well known that resonant tunneling effects observed in quantum-dot structures strongly depend on an applied bias voltage and the electronic structure of the quantum dot. For understanding detailed electron-transport properties of nanostructures, it is essential and indispensable to take into account the effects of applied bias voltages and electric fields on electron conduction through nanostructures.

Let us consider a system that consists of only bare left and right semi-infinite jellium electrodes in the absence of the atom, as shown in Fig. 10.1(a). The two jellium electrodes are connected through a power supply providing a bias voltage of V. Then, the Fermi level and effective potential of the left electrode are negatively boosted up by the energy corresponding to the bias voltage V relatively to the right electrode. The difference between the Fermi levels, V, seems to be quite the same as that between the effective potentials, W, when the two jellium electrodes are identical. However, in carrying out self-consistent calculations for the

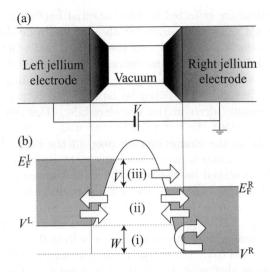

Fig. 10.1 (a) Schematic representation of a system of two semi-infinite jellium electrodes facing to each other with a finite separation and (b) its energetic diagram. A finite bias voltage is applied to the bare electrode system in the absence of the atom. The Fermi levels and effective potentials inside the electrodes are represented by $E_F^{L(R)}$ and $V^{L(R)}$, respectively.

charge density and potential, we cannot regard these two values, W and V, as agreeing exactly with each other. We shall discuss below the reason why this discrepancy is led [Lang (1992); Di Ventra and Lang (2001)].

When the jellium electrodes are *ideal* electron reservoirs with the characteristic (iii) mentioned in Section 6.1, valence electrons inside the left (right) electrode must continuously occupy the electronic states within the energy band from the effective potential V^L (V^R) to the Fermi level E_F^L (E_F^R), as seen in Fig. 10.1(b). It is convenient to define three energy ranges: The first is the range (i) between the effective potentials of the right and left electrodes, in which there are no electronic states of valence electrons in the left electrode and those in the right one are fully occupied by valence electrons. The next is the range (ii) from the effective potential of the left electrode to the Fermi level of the right electrode, where valence electrons exist in both the electrodes. The last is the range (iii) between the Fermi levels of the two electrodes, in which electronic states in the left electrode are fully occupied by valence electrons and those in the right one are completely empty.

Possible electron injections and scatterings simply imagined are as follows. In the lowest energy range (i), all electrons that are injected from the right electrode are reflected by the potential barrier at a vacuum region, and are not allowed to arrive at the left electrode due to absence of electronic states. Consequently, all the injected and reflected electrons contribute to the formation of the charge density in the right electrode. For the middle energy range (ii), a part of electrons coming from the left electrode are reflected by the potential barrier, and return to contribute to the charge density deep in the left electrode. The rest of electrons, which have transmitted the potential barrier with a transmission probability, contribute to the charge density deep in the right electrode. On the other hand, electrons injected from the right electrode are also able to transmit the potential barrier with the same transmission probability as that of electrons injected from the left electrode. Therefore, the total amounts of electrons in the left and right electrodes within the energy range (ii) do not change. Let us proceed to the highest energy range (iii). From Fig. 10.1(b), we recognize electrons coming only from the left electrode, because in the right electrode the energy range (iii) corresponds to the empty states. This means that only the electrons injected from the left electrode contribute to the formation of the charge density in the both electrodes, namely, transmitted electrons contribute to the charge density in the right electrode, and reflected electrons contribute to that in the left electrode. Resultantly, the charge density deep in the left (right) electrode would become slightly smaller (larger) than the original charge density deep in the jellium electrode.

Self-consistent calculations require that the electron charge density neutralizes the positive background deep inside the electrodes. Thus, we have to adjust the energy-band width $(E_F^{L(R)} - V^{L(R)})$ of each jellium electrode. The Fermi levels of both the electrodes must be fixed so as to have a definite constant difference V according to a constrained bias voltage of V. Therefore, the bottoms of the energy bands (effective potentials) of both the electrodes turn out to be changed to cancel the charge unbalance; in Fig. 10.1(b), the energy-band width of the left electrode, $E_F^L - V_L$, should increase and that of the right electrode, $E_F^R - V_R$, should decrease. This adjustment of the energy-band widths might be negligible for a system with a large electrode separation, because this effect is strongly attributed to the electron transmitting probability. But, when we discuss a system where both the electrodes have a smaller separation than 1 nm or are connected via nanostructures, this effect must be taken into account in self-consistent calculations of the charge density and potential. As an example, at a bias voltage of 6 V, the difference between effective potentials, W, becomes 5.96 V for an electrode separation of 8.5 a.u. [Lang (1992)].

10.3 Application: Sodium Nanowire under Finite Bias Voltages

In this section, we present the electric properties of a sodium monoatomic wire, which are evaluated by the method based on the Lippmann–Schwinger equation. All the calculations for determining the electronic structure and transport properties are carried out self-consistently, and the norm-conserving pseudopotential technique [Troullier and Martins (1991)] is implemented to describe the electron-ion interaction correctly. The sodium nanowire sandwiched between two metallic electrodes is found to exhibit a unique I–V characteristic, such as a negative differential conductance (NDC) behavior. Moreover, it is observed that voltage drop is not generated uniformly along the nanowire axis but is mainly induced at the negatively biased end of the nanowire. We will show below that this strongly localized voltage drop is understood in consideration of the geometry of the nanowire system and electron flow, and that a detailed analysis of electronic structures of the nanowire system including the voltage drop distribution reveals the mechanism of the occurrence of the NDC behavior.

The schematic of the sodium nanowire system employed for the calculation of the electric properties is drawn in Fig. 10.2. The nanowire system is composed of the following three components: The first is the monoatomic straight wire consisting of five sodium atoms. The length of this straight nanowire is 28.0 a.u., which is equal to four times the nearest-neighbor atomic distance $d = 7.0$ a.u. The second is a couple of square bases with a side length of 8.2 a.u., each of which is constructed by four sodium atoms. The two square bases are attached to both ends of the straight nanowire, and one basis and one end atom of the straight nanowire form a tetrag-

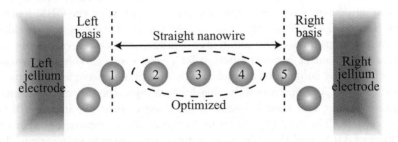

Fig. 10.2 Schematic of model of sodium nanowire system. The balls represent the atomic positions. Left- and right-hand-side solids are the semi-infinite jellium electrodes. All the atoms except those enclosed by a broken circle are fixed at the initial positions during geometrical optimization.

onal pyramid with a height of 4.0 a.u. The last is a pair of semi-infinite jellium electrodes, where the electron density is taken to be equal to the average valence-electron density of the sodium crystal ($r_s = 4.0$ a.u.). The atomic structure composed of a straight nanowire and a couple of bases is sandwiched between the pair of jellium electrodes. To moderate possible unphysical effects due to an oversimplified electrode model of jellium, the two square bases can work as buffer layers.

Using the real-space finite-difference method developed in Part I, the sodium atoms constituting the straight nanowire have been optimized geometrically under the following three restrictions: (i) The geometrical optimization is carried out without application of any finite electric field. (ii) The length of the straight nanowire is fixed to be $4d$. (iii) Both of the end atoms of the straight nanowire are kept frozen. The geometrical relaxation of the straight nanowire is continued until the remaining forces acting on the nuclei are less than 0.005 Ry/Å. Then, the relaxed atoms form a straight line so as to trimerize in the central region with a length of 13.9 a.u. The electronic structures of the sodium nanowire system, upon the application of sizable electric fields, are self-consistently determined under the optimized geometrical parameters. In order to generate an electric field between the electrodes, the Fermi level of the left electrode is moved so that the bias voltage is applied with the left electrode negative, while the Fermi level of the right electrode is kept fixed.

Figure 10.3 shows the I–V characteristic and its differential conductance. The current is simply expressed as the integration of the electron transmission over the voltage window. The voltage window is defined as the energy range between the Fermi levels of the two electrodes (see Fig. 10.6 for illustration). The differential conductance is defined as the derivative of the current flow with respect to the bias voltage. The quantized differential conductance of 1 G_0 appears for the infinitesimal bias voltage. This is in excellent agreement with the calculation result obtained by the OBM method (see Section 6.6). When the bias voltage applied to the system is raised up to 1.0 V gradually, the differential conductance decreases rapidly and seems to remain at approximately 0.7 G_0. As the bias voltage is raised, the differential conductance quickly falls and finally becomes negative at about 2.1 V. At approximately this bias voltage, the current flow has a maximal value, and then it begins to decrease against the increase of the bias voltage. The current flow turns to rise again at approximately a bias voltage of 3.0 V, and the differential conductance turns positive again. The NDC behavior of the differential conductance is clearly recognized at bias voltages of 2.1 – 3.0 V. The unique NDC behavior is expected to be an essential property that allows fast switching in certain types of future electronic devices. For research studies concerning NDC, see, for example, Lyo

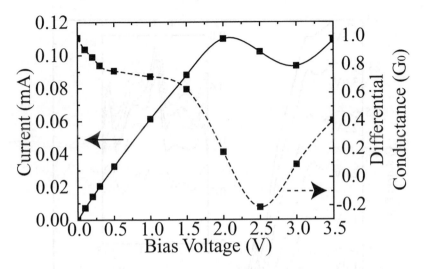

Fig. 10.3 I–V characteristic (solid curve) and differential conductance (dashed curve) of the sodium nanowire system illustrated in Fig 10.2. Reprinted with permission from Tsukamoto and Hirose (2002). © 2002 American Physical Society.

and Avouris (1989), Lang (1997c), Di Ventra *et al.* (2001) and Tsukamoto *et al.* (2002).

We next examine how the applied bias voltage drops along the sodium nanowire axis. The voltage drop along the wire is one important open problem in nanoelectronics [Lang and Avouris (2000)]. Figure 10.4 illustrates the differences of the local effective potentials along the axis at some significant bias voltages from that at a zero bias voltage, $V_{\text{eff}}(V_B) - V_{\text{eff}}(0.0)$. These differences are considered to indicate voltage drops along the nanowire axis. In Fig. 10.4, one sees that as the bias voltage applied to the left electrode increases negatively, the potential around the left basis rises in proportion to the absolute value of the applied bias voltage; on the other hand, the potential around the right basis hardly shifts. Therefore, for any bias voltage, the electric potential of each basis is equal to that of the jellium electrode to which the basis is attached. This seems consistent with electromagnetism, i.e., the fact that no electric fields are generated inside electric conductors. The voltage drop between the left and right electrodes is observed to take place mainly around the negatively biased end of the nanowire (atoms 1 and 2 in Fig. 10.4), while around the right half of the straight nanowire (atoms numbered 3 – 5), the potential drop is found to be relatively small. This unique voltage-drop distribution along the nanowire axis conflicts with Ohm's law for diffusive electronic trans-

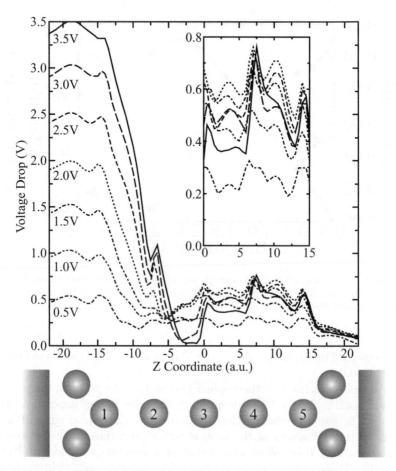

Fig. 10.4 Shifts of the local effective potential along the axis of the sodium nanowire upon applying finite bias voltages. Vertically magnified figure is depicted on the upper right panel. The atomic geometry of the system is drawn below as visual guides. Some of data taken from Tsukamoto and Hirose (2002).

port, which expects the emergence of a uniform electric field. The strong localization of the voltage drop is understood as follows: Because of the lowering of the dimension for electrons flowing from left to right, i.e., the geometrical transition of the electron path from a three-dimensional wide electrode to a one-dimensional narrow nanowire, electrons injected from the negative electrode are focused and immediately introduced into the straight nanowire around the region between atoms 1 and 2. Although only electrons with some specific momenta can enter conduction channels and reach

$\rho(3.0\text{V})-\rho(0.0\text{V})$

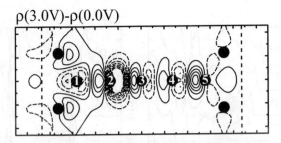

Fig. 10.5   Induced electron-density distribution due to an applied bias voltage of 3.0 V. The change in the charge density is depicted on the cross section including the nanowire axis and some bases atoms. The solid lines represent the electron excess, and the broken lines denote the electron depletion. The vertical broken lines indicate the edges of the jellium electrodes. Reprinted with permission from Tsukamoto and Hirose (2002). © 2002 American Physical Society.

the positive electrode without backscattering, the remaining electrons suffer from a considerable amount of reflection. Thus, the resistance for electron flow localizes between atoms 1 and 2, which results in the intensive voltage drop there. This mechanism of electron flow is analogous to the *vena contracta* effect known in hydrodynamics. A similar feature of the voltage drop has been observed for gold nanowires [Brandbyge *et al.* (1999)] and other nanowires [Hirose *et al.* (2004)].

The absence of a sizable voltage drop near the two bases and on the positively biased side of the nanowire is associated with the change in the charge density distribution upon applying bias voltage. Figure 10.5 shows the profile of the difference of the charge densities, $\rho(3.0 \text{ V}) - \rho(0.0 \text{ V})$. On the right of the atoms forming the left basis and the left of atom 2, one observes considerable increases of the charge density. These increases reflect the screening effect which keeps the electric potential of the negatively biased jellium electrode and basis equal to the negative electrode potential applied to the left jellium. On the other hand, the most significant decreases of electrons occur on the right of atom 2 so that the electric potential of the positively biased side of the nanowire system (atoms $3-5$, the right basis and jellium electrode) becomes close to the positive electrode potential assigned to the right jellium. As a consequence of the negative electric potential on the left of atom 1 and the positive one on the right of atom 2, the localization of the voltage drop between atoms 1 and 2 is naturally realized, as seen in Fig. 10.4.

In order to investigate the nature of the NDC behavior observed in Fig. 10.3, we pay attention to the total electron transmissions and LDOS around the straight nanowire. Figure 10.6(a) shows the electron transmis-

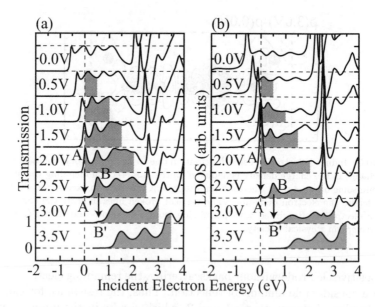

Fig. 10.6 Electron transmission (a) and LDOS around the straight nanowires (b) drawn as a function of the energy of electrons injected from the left electrode. The electron energy is measured from the Fermi level of the right electrode. The applied bias voltages are shown in the panels. Shaded areas represent the voltage windows. The curves are displaced vertically for clarity. Reprinted with permission from Tsukamoto and Hirose (2002). © 2002 American Physical Society.

sions for several bias voltages. The vertical broken line corresponds to the Fermi level of the right electrode, and the right-hand edges of the voltage windows (shaded areas) correspond to the Fermi level of the bias-applied left electrode. Electrons belonging to this area contribute to current flowing through the system. Figure 10.6(b) displays the LDOS around the straight nanowire for several bias voltages. Here, these LDOS curves are derived from the wave functions of electrons injected only from the left electrode. One can see that as the bias voltage increases from 0.0 V to 1.0 V, the entire electron transmission curve moves to the high-energy side by about 0.5 eV. This movement reflects the increase of the electric potential around the straight nanowire under the application of a bias voltage of 1.0 V (see Fig. 10.4). For bias voltages higher than 1.5 V, the electron transmission curve does not seem to move any more. This is easily verified by tracing a peak position, which is found at 2.0 eV in the transmission curve at a zero bias voltage. The pinning of the transmission curve is associated with the small change in the electric potential around the straight nanowire against rising the bias voltage (see Fig. 10.4). The most interesting is the sudden suppression of electron-transmission peaks, i.e., the leftmost peak seen at the bias voltage of 2.0 V is suddenly suppressed at 2.5 V (A→A'). The left peak at 2.5 V, indicated by B in Fig. 10.6(a), is also suppressed when the bias voltage is raised to 3.0 V (B→B'). As a consequence of the peak suppression, the integration of the electron transmission over the voltage window, namely the current flowing through the system, decreases upon raising the bias voltage from 2.0 V to 3.0 V, which leads to the NDC behavior. In contrast, such disappearance of electron-transmission peaks is no longer observed at the bias voltage of 3.5 V, and then, electron flow expectedly increases as the bias voltage is boosted up from 3.0 V to 3.5 V, as was shown in Fig. 10.3.

The sudden suppression of the electron-transmission peaks seems to go hand in hand with the LDOS around the straight nanowire for the corresponding bias voltages, because the sudden vanishing of LDOS peaks, marked by A→A' and B→B' in Fig. 10.6(b), is observed at the same position and at the same bias voltage where the electron-transmission peaks are suddenly suppressed. This vanishing of LDOS peaks implies that when the bias voltage exceeds a certain threshold, some electrons coming from the left electrode with specific energies cannot reach the straight nanowire, i.e., the electrons must be blocked and reflected by the left basis.

In order to clarify the reason why the electron-transmission peaks and LDOS peaks suddenly disappear when the bias voltage exceeds some certain threshold voltages, as observed in Fig. 10.6, let us extend our view to the left and right bases, and investigate the LDOS of electrons injected from the left electrode. Figures 10.7(a) and (b) show the LDOS for bias voltages of 0.0 – 3.5 V around the left and right bases, respectively. We can recognize two features in the series of LDOS curves for the right basis [Fig. 10.7(b)], which are similar to those in the series of LDOS curves for the straight nanowire [Fig. 10.6(b)]: One is that the LDOS curve moves to the high-energy side by 0.5 eV at bias voltages up to 1.0 V, and seems to remain constant at bias voltages of 1.5 V and above. Another is that the LDOS peaks observed clearly at 0.0 eV and 0.5 eV in Fig. 10.7(b) suddenly vanish, as indicated by A→A' and B→B', respectively. On the other hand, the series of LDOS curves for the left basis [Fig. 10.7(a)] exhibits quite different behavior from that for the right basis: All the LDOS curves for the left basis just shift to the high-energy side in proportion to the absolute value of the applied bias voltage while maintaining their shapes, which is in marked contrast to the LDOS around the straight nanowire [see Fig. 10.6(b)]. This uniform shift of the LDOS around the left basis reflects the linear rising of the electric potential around the left basis upon applying bias voltage (see Fig. 10.4).

In Fig. 10.7(a), we can recognize that the leftmost LDOS peaks for the

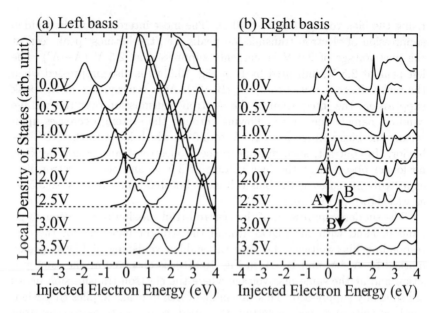

Fig. 10.7   LDOS around the left (a) and right (b) bases. The meanings of the other parameters are the same as those in Fig. 10.6. Some of data taken from Tsukamoto and Hirose (2002).

left basis stand at 0.0, 0.5 and 1.0 eV at bias voltages of 2.0, 2.5 and 3.0 V, respectively. These LDOS peaks do play an important role in understanding the sudden suppression of the electron-transmission peaks, namely, the origin of the NDC phenomenon. Now, let us clarify the mechanism of the NDC observed in the sodium nanowire system. In general, the energy at which the leftmost LDOS peak is observed coincides with the first onset energy in an electron-transmission curve [Kobayashi and Tsukada (1999); Tsukamoto *et al.* (2002)]. Thus, the appearance of the first peak at 0.0 eV in the curve of LDOS around the left basis at the bias voltage of 2.0 V indicates that electrons with the energies lower than 0.0 eV are refused to go through the left basis. On the other hand, electrons with energies of 0.0 eV and above are readily able to pass through the left basis, and enter the straight nanowire to yield the LDOS peak indicated by A in Fig. 10.6(b). The schematic energy diagram and electron-transport trajectory are depicted in Fig. 10.8(b). Therefore, the electron-transmission peak of A in Fig. 10.6(a) is also observed and contributes to the current flowing through the entire nanowire system. At the bias voltage of 2.5 V, it is now easily understood that the threshold energy for electrons to go through the left basis is boosted up to be 0.5 eV, because the leftmost peak in the LDOS curve for the left basis stands at 0.5 eV. Hence, electrons with an energy of 0.0 eV are prohibited from entering the left basis, and cannot reach the straight nanowire. This electron reflection procedure is schematically drawn in Fig. 10.8(c). As the result, the LDOS peak for the straight nanowire at 0.0 eV disappears at the bias voltage of 2.5 V. Thus, when the bias voltage is boosted from 2.0 V to 2.5 V, the vanishing of the LDOS peak around the straight nanowire at 0.0 eV and the sudden suppression of the transmission peak at 0.0 eV, A→A', are induced. This mechanism of the suppression of the transmission peak also holds in the case of B→B' in Fig. 10.6(a). Further details of the study are presented in Tsukamoto and Hirose (2002) and elsewhere.

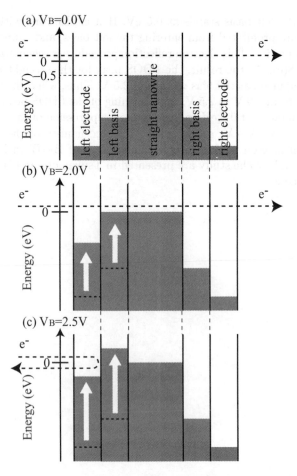

Fig. 10.8   Schematic representation of the mechanism of NDC observed in the sodium nanowire system. V_B indicates the applied bias voltage. The energy diagram at a zero bias voltage is depicted in (a). The dark areas are the energy range where electronic transport is not allowed. Then, in the case of (a), the straight nanowire region dominates the electronic transport through the system, and the electron-transmission curve has an onset at -0.5 eV. When the applied bias voltage rises up to 2.0 V (b), the dark areas of the left basis and straight nanowire rise up according to the electric potential shown in Fig. 10.4, and electrons with an energy of 0.0 eV can pass through the system barely. In the case of the applied bias voltage of 2.5 V (c), the dark area of the left-basis region is boosted up to overtake that of the straight-nanowire region due to the rising of the electric potential around the left basis. Then, the left basis becomes dominant in the electronic transport through the system, and reflects electrons with an energy 0.0 eV. Therefore, the electronic transport contributing to the current flow for the lower bias voltage suddenly turns off at the bias voltage of 2.5 V.

Formulas for Long-Range Potentials under Various Boundary Conditions

Tomoya Ono

Computing Coulomb potentials and energies in infinite systems requires much computational cost and involves numerical difficulties, since the potential of $1/|\boldsymbol{r}|$ slowly vanishes at the limit of $|\boldsymbol{r}| \to \infty$. In this appendix, an efficient method proposed by Ewald (1921) for treating the sums of Coulomb potentials and energies is explained. Not employing fast Fourier transforms, this method is suitable for parallel computing. Although the formulas in the cases using the pseudopotentials of Bachelet *et al.* [Hamann *et al.* (1979); Bachelet *et al.* (1982)] as local components are described here, this method is applicable to other types of pseudopotentials.

A.1 3D Periodic Boundary Condition

Ionic pseudopotential

The ionic pseudopotential is given by

$$V_{loc}^s(\boldsymbol{r}) = \sum_{\boldsymbol{P}} v_{loc}^s(\boldsymbol{P} + \boldsymbol{r} - \boldsymbol{R}^s), \tag{A.1}$$

where

$$v_{loc}^s(\boldsymbol{r}) = -\frac{Z_s}{|\boldsymbol{r}|} \sum_{i=1,2} C_{s,i}\, \mathrm{erf}(\sqrt{\alpha_{s,i}}\,|\boldsymbol{r}|), \tag{A.2}$$

the sum is over real-space lattice vectors $\boldsymbol{P}$ of $(n_x L_x, n_y L_y, n_z L_z)$, and L_x, L_y, and L_z are the lengths of the unit cell in the x, y, and z directions, respectively. In addition, $C_{s,i}$ $(C_{s,1} + C_{s,2} = 1)$ and $\alpha_{s,i}$ are the parameters of the pseudopotential given by Bachelet *et al.* [Hamann *et al.* (1979); Bachelet *et al.* (1982)], and $\mathrm{erf}(x)$ is the error function (or probability integral) defined by

$$\text{erf}(x) = \frac{2}{\sqrt{\pi}} \int_0^x e^{-t^2} dt. \tag{A.3}$$

Substituting (A.2) into (A.1), we obtain

$$V_{loc}^s(\boldsymbol{r}) = -Z_s \sum_{\boldsymbol{G}} \frac{2\pi}{\Omega} \exp\left[i\boldsymbol{G} \cdot (\boldsymbol{r} - \boldsymbol{R}^s)\right] \int_0^{\eta} \exp\left(-\frac{|\boldsymbol{G}|^2}{4t^2}\right) \frac{1}{t^3} dt$$

$$-Z_s \sum_{\boldsymbol{P}} \frac{2}{\sqrt{\pi}} \sum_{i=1,2} C_{s,i} \int_{\eta}^{\sqrt{\alpha_{s,i}}} \exp(-|\boldsymbol{P} + \boldsymbol{r} - \boldsymbol{R}^s|^2 t^2) dt, \tag{A.4}$$

where $\boldsymbol{G}$ are reciprocal lattice-vectors of $2\pi(\frac{j_x}{L_x}, \frac{j_y}{L_y}, \frac{j_z}{L_z})$ and Ω is a volume of the unit cell. This identity of the Ewald prescription provides a method for rewriting the lattice summation for the Coulomb potential at $\boldsymbol{r}$ due to the contribution from an array of atoms positioned at the points $\boldsymbol{R}^s - \boldsymbol{P}$. The Coulomb interactions are partitioned into a long-range contribution summed in reciprocal space and a short-range sum handled in real space. When η is chosen to be $0.2 - 0.7$, the amount of the sums in (A.4) is only $7^3 - 11^3$ operations.

In a charge-neutral system, the contributions to the total potential from the ionic pseudopotentials and Hartree potential at $\boldsymbol{G} = 0$ cancel exactly if the pseudopotential is a pure Coulomb potential, and so the $\boldsymbol{G} = 0$ contribution to the Coulomb potential of the ionic system must be removed in order to compute the correct potential. The local component of the pseudopotential proposed by Bachelet *et al.* [Hamann *et al.* (1979); Bachelet *et al.* (1982)], however, is not a pure potential, and there is a constant contribution to Coulomb potential at $\boldsymbol{G} = 0$, equal to the integral of the difference between the pure Coulomb potential and pseudopotential. This constant per volume is

$$v_{diff}^s = \frac{\pi}{\Omega} Z_s \left(\frac{C_{s,1}}{\alpha_{s,1}} + \frac{C_{s,2}}{\alpha_{s,2}}\right). \tag{A.5}$$

Moreover, in the Ewald summations, the $\boldsymbol{G} = 0$ contribution to the Coulomb potential has been divided between the real-space and reciprocal-space summations, so that it is not sufficient simply to omit the $\boldsymbol{G} = 0$ term in the reciprocal-space Ewald summation. The $\boldsymbol{G} = 0$ term in the reciprocal-space summation should be omitted with the addition of one term $\pi Z_s / \Omega \eta^2$ to the Ewald potential to give the correct potential. The correct form for the ionic pseudopotential is

$$
V_{loc}^s(r) = -\frac{4\pi}{\Omega} Z_s \sum_{G \neq 0} \frac{1}{|G|^2} \cos\left[G \cdot (r - R^s)\right] \exp\left(-\frac{|G|^2}{4\eta^2}\right) + \frac{\pi}{\Omega} \frac{Z_s}{\eta^2}
$$

$$
+ Z_s \sum_{P} \frac{1}{|\zeta^s|} \left[\operatorname{erf}(\eta\,|\zeta^s|) - \sum_{i=1,2} C_{s,i}\,\operatorname{erf}(\sqrt{\alpha_{s,i}}\,|\zeta^s|)\right],
$$

$$(A.6)$$

where $\zeta^s = P + r - R^s$ and the sum of $\sum_{G \neq 0}$ is over the reciprocal lattice vectors excluding the case of $|G| = 0$. In addition, $V_{loc}^s(r)$ holds

$$
\int_\Omega V_{loc}^s(r)dr = \Omega\,v_{diff}^s. \tag{A.7}
$$

Hartree potential

In the case of solving the Poisson equation of (2.8) under the periodic boundary condition in the x, y, and z directions, the constant contribution such as the $G = 0$ component in the ionic pseudopotential to the Hartree potential v_H is required instead of determining the boundary values. As mentioned above, it cancels exactly, and therefore one has only to solve the Poisson equation under the constraint of

$$
\int_\Omega v_H(r)dr = 0. \tag{A.8}
$$

Coulomb energy among the nuclei

There are similar divergences in the Coulomb energy among the nuclei. The energy can be evaluated in the same manner as the ionic pseudopotential. The resultant energy is as follows:

$$
\gamma_E = \frac{2\pi}{\Omega} \sum_{s,s'} Z_s Z_{s'} \sum_{G \neq 0} \frac{1}{|G|^2} \cos\left[G \cdot (R^{s'} - R^s)\right] \exp\left(-\frac{|G|^2}{4\eta^2}\right)
$$

$$
- \sum_{s,s'} Z_s Z_{s'} \left(\frac{\pi}{2\eta^2\Omega} + \delta_{ss'}\frac{\eta}{\sqrt{\pi}}\right) + \frac{1}{2} \sideset{}{'}\sum_{P,s,s'} Z_s Z_{s'} \frac{\operatorname{erfc}(\eta \cdot |\xi^{s,s'}|)}{|\xi^{s,s'}|},
$$

$$(A.9)$$

where $\xi^{s,s'} = P + R^{s'} - R^s$, and $\operatorname{erfc}(x)$ is the complementary error function:

$$
\operatorname{erfc}(x) = 1 - \operatorname{erf}(x). \tag{A.10}
$$

The differentiation of γ_E with respect to the position of the s-th nucleus $\boldsymbol{R}^s$ is as follows: For $\mu = x$, y, and z,

$$-\frac{\partial \gamma_E}{\partial R^s_\mu} = -\frac{2\pi}{\Omega} \sum_{s'} Z_s Z_{s'} \sum_{\boldsymbol{G} \neq 0} \frac{G_\mu}{|\boldsymbol{G}|^2} \sin\left[\boldsymbol{G} \cdot (\boldsymbol{R}^{s'} - \boldsymbol{R}^s)\right] \exp\left(-\frac{|\boldsymbol{G}|^2}{4\eta^2}\right)$$

$$-\sum_{\boldsymbol{P},s'}' Z_s Z_{s'} \left[\frac{\eta \exp(-\eta^2 |\boldsymbol{\xi}^{s,s'}|^2)}{\sqrt{\pi}|\boldsymbol{\xi}^{s,s'}|} + \frac{\mathrm{erfc}(\eta |\boldsymbol{\xi}^{s,s'}|)}{2|\boldsymbol{\xi}^{s,s'}|^2}\right] \frac{\xi^{s,s'}_\mu}{|\boldsymbol{\xi}^{s,s'}|}.$$

$$\text{(A.11)}$$

A nucleus does not interact with its own Coulomb charge, so that the $\boldsymbol{P} = 0$ term must be omitted from the real-space summation when $s = s'$. The prime in the last summation in (A.9) and (A.11) means that $|\boldsymbol{\xi}^{s,s'}| = 0$ is omitted.

A.2 (2D Periodic + 1D Isolated) Boundary Condition

Similarly to the case of the 3D periodic boundary condition, the potential and energy in the case of a (2D periodic + 1D isolated) boundary condition are obtained by the Ewald summation. Here, the case in which periodic boundary conditions are imposed in the x and y directions and an isolated boundary condition in the z direction is explained.

Ionic pseudopotential

By using the identity $(b > 0)$,

$$\int_0^1 \frac{1}{t^2} \exp\left(-a^2 t^2 - \frac{b^2}{4t^2}\right) dt$$

$$= \frac{\sqrt{\pi}}{2b} \left[\exp(-ab)\,\mathrm{erfc}\left(\frac{b-2a}{2}\right) + \exp(ab)\,\mathrm{erfc}\left(\frac{b+2a}{2}\right)\right], \text{ (A.12)}$$

the ionic pseudopotential is given by

$$V^s_{loc}(\boldsymbol{r}) = -\frac{\pi}{S} Z_s \sum_{|\boldsymbol{G}| \neq 0} \frac{1}{|\boldsymbol{G}|} \cos\left[\boldsymbol{G} \cdot (\boldsymbol{r} - \boldsymbol{R}^s)\right] f^+(\boldsymbol{G}, \boldsymbol{r})$$

$$+\frac{2\sqrt{\pi}}{S} Z_s \left[\frac{1}{\eta} \exp(-|z - R^s_z|^2 \eta^2) + \sqrt{\pi}\,|z - R^s_z|\,\mathrm{erf}(|z - R^s_z|\,\eta)\right]$$

$$+Z_s \sum_{\boldsymbol{P}} \frac{1}{|\boldsymbol{\zeta}^s|} \left[\mathrm{erf}(\eta\,|\boldsymbol{\zeta}^s|) - \sum_{i=1,2} C_{s,i}\,\mathrm{erf}(\sqrt{\alpha_{s,i}}\,|\boldsymbol{\zeta}^s|)\right], \qquad \text{(A.13)}$$

where $G = 2\pi(\frac{j_x}{L_x}, \frac{j_y}{L_y}, 0)$, $P = (n_x L_x, n_y L_y, 0)$, $S = L_x \times L_y$, $\zeta^s = P + r - R^s$, and

$$f^{\pm}(G, r) = \exp\left[-|G|\,(z - R_z^s)\right] \mathrm{erfc}\left(\frac{|G| - 2\eta^2(z - R_z^s)}{2\eta}\right)$$
$$\pm \exp\left[|G|\,(z - R_z^s)\right] \mathrm{erfc}\left(\frac{|G| + 2\eta^2(z - R_z^s)}{2\eta}\right).$$

$$(A.14)$$

Hartree potential

To solve the Poisson equation under the combination of periodic and isolated boundary conditions, the boundary values of the direction on which an isolated boundary condition is imposed are required, and they are evaluated according to (2.21). Here, the Ewald−Kohnfeld prescription for computing the second and third term of (2.21) is explained. We first determine

$$V_H^{1,s}(r) \equiv \sum_P \frac{1}{|\zeta^s|},$$

$$(A.15)$$

$$V_{H,\mu}^{2,s}(r) \equiv \sum_P \frac{\zeta_\mu^s}{|\zeta^s|^3} = \frac{\partial}{\partial R_\mu^s} V_H^{1,s}(r),$$

$$(A.16)$$

and

$$V_{H,\mu\nu}^{3,s}(r) \equiv \sum_P \frac{3\,\zeta_\mu^s \zeta_\nu^s - \delta_{\mu\nu}|\zeta^s|^2}{|\zeta^s|^5} = \frac{\partial}{\partial R_\mu^s} \frac{\partial}{\partial R_\nu^s} V_H^{1,s}(r).$$

$$(A.17)$$

Thus, (2.21) is written as

$$v_H(r) = \sum_s \left(\int \rho_s(r')dr' \cdot V_H^{1,s}(r) + \sum_{\mu = x,y,z} p_{s,\mu} \cdot V_{H,\mu}^{2,s}(r) \right.$$
$$\left. + \sum_{\mu,\nu = x,y,z} q_{s,\mu\nu} \cdot V_{H,\mu\nu}^{3,s}(r) + \cdots \right).$$

$$(A.18)$$

Here, it is easily recognized that

$$V_H^{1,s}(r) = -\lim_{\alpha_{s,i} \to \infty} \frac{V_{loc}^s(r)}{Z_s},$$

$$(A.19)$$

and therefore,

$$V_H^{1,s}(\boldsymbol{r}) = \frac{\pi}{S} \sum_{|\boldsymbol{G}| \neq 0} \frac{1}{|\boldsymbol{G}|} \cos\left[\boldsymbol{G} \cdot (\boldsymbol{r} - \boldsymbol{R}^s)\right] f^+(\boldsymbol{G}, \boldsymbol{r})$$

$$- \frac{2\sqrt{\pi}}{S} \left[\frac{1}{\eta} \exp(-|z - R_z^s|^2 \, \eta^2) + \sqrt{\pi}|z - R_z^s| \operatorname{erf}(|z - R_z^s| \, \eta)\right]$$

$$+ \sum_{\boldsymbol{P}} \frac{1}{|\boldsymbol{\zeta}^s|} \operatorname{erfc}(\eta \, |\boldsymbol{\zeta}^s|). \tag{A.20}$$

$V_{H,\mu}^{2,s}(\boldsymbol{r})$ are as follows: in the case of $\mu = x, y$,

$$V_{H,\mu}^{2,s}(\boldsymbol{r}) = \frac{\pi}{S} \sum_{|\boldsymbol{G}| \neq 0} \frac{G_\mu}{|\boldsymbol{G}|} \sin\left[\boldsymbol{G} \cdot (\boldsymbol{r} - \boldsymbol{R}^s)\right] f^+(\boldsymbol{G}, \boldsymbol{r})$$

$$+ \sum_{\boldsymbol{P}} \left[\frac{2\eta \exp(-\eta^2 \, |\boldsymbol{\zeta}^s|^2)}{\sqrt{\pi} \, |\boldsymbol{\zeta}^s|} + \frac{\operatorname{erfc}(\eta \, |\boldsymbol{\zeta}^s|)}{|\boldsymbol{\zeta}^s|^2}\right] \frac{\zeta_\mu^s}{|\boldsymbol{\zeta}^s|}, \tag{A.21}$$

and in the case of $\mu = z$,

$$V_{H,\mu}^{2,s}(\boldsymbol{r}) = \frac{\pi}{S} \sum_{|\boldsymbol{G}| \neq 0} \cos\left[\boldsymbol{G} \cdot (\boldsymbol{r} - \boldsymbol{R}^s)\right] f^-(\boldsymbol{G}, \boldsymbol{r})$$

$$+ \frac{2\pi}{S} \operatorname{erf}(|z - R_z^s| \, \eta) \frac{(z - R_z^s)}{|z - R_z^s|}$$

$$+ \sum_{\boldsymbol{P}} \left[\frac{2\eta \exp(-\eta^2 \, |\boldsymbol{\zeta}^s|^2)}{\sqrt{\pi} \, |\boldsymbol{\zeta}^s|} + \frac{\operatorname{erfc}(\eta \, |\boldsymbol{\zeta}^s|)}{|\boldsymbol{\zeta}^s|^2}\right] \frac{\zeta_\mu^s}{|\boldsymbol{\zeta}^s|}. \tag{A.22}$$

$V_{H,\mu\nu}^{3,s}(\boldsymbol{r})$ are as follows: in the case of $\mu = x, y$ and $\nu = x, y$,

$$V_{H,\mu\nu}^{3,s}(\boldsymbol{r}) = -\frac{\pi}{S} \sum_{|\boldsymbol{G}| \neq 0} \frac{G_\mu G_\nu}{|\boldsymbol{G}|} \cos\left[\boldsymbol{G} \cdot (\boldsymbol{r} - \boldsymbol{R}^s)\right] f^+(\boldsymbol{G}, \boldsymbol{r})$$

$$+ \sum_{\boldsymbol{P}} \left[\left\{\frac{4\eta \exp(-\eta^2|\boldsymbol{\zeta}^s|^2)}{\sqrt{\pi}} \left(\eta^2 + |\boldsymbol{\zeta}^s|^{-2}\right) + \frac{2\operatorname{erfc}(\eta|\boldsymbol{\zeta}^s|)}{|\boldsymbol{\zeta}^s|^3}\right\} \frac{\zeta_\mu^s \zeta_\nu^s}{|\boldsymbol{\zeta}^s|^2}\right.$$

$$\left. + \left\{\frac{2\eta \exp(-\eta^2 \, |\boldsymbol{\zeta}^s|^2)}{\sqrt{\pi} \, |\boldsymbol{\zeta}^s|} + \frac{\operatorname{erfc}(\eta \, |\boldsymbol{\zeta}^s|)}{|\boldsymbol{\zeta}^s|^2}\right\} \frac{\zeta_\mu^s \zeta_\nu^s - \delta_{\mu\nu}|\boldsymbol{\zeta}^s|^2}{|\boldsymbol{\zeta}^s|^3}\right], \tag{A.23}$$

in the case of $\mu = x, y$ and $\nu = z$,

$$V_{H,\mu\nu}^{3,s}(\boldsymbol{r}) = \frac{\pi}{S} \sum_{|\boldsymbol{G}|\neq 0} G_\mu \sin\left[\boldsymbol{G}\cdot(\boldsymbol{r}-\boldsymbol{R}^s)\right] f^-(\boldsymbol{G},\boldsymbol{r})$$

$$+ \sum_{P}\left[\left\{\frac{4\eta\exp(-\eta^2|\boldsymbol{\zeta}^s|^2)}{\sqrt{\pi}}\left(\eta^2+|\boldsymbol{\zeta}^s|^{-2}\right)+\frac{2\,\mathrm{erfc}(\eta|\boldsymbol{\zeta}^s|)}{|\boldsymbol{\zeta}^s|^3}\right\}\frac{\zeta_\mu^s\zeta_\nu^s}{|\boldsymbol{\zeta}^s|^2}\right.$$

$$\left.+\left\{\frac{2\eta\exp(-\eta^2|\boldsymbol{\zeta}^s|^2)}{\sqrt{\pi}|\boldsymbol{\zeta}^s|}+\frac{\mathrm{erfc}(\eta|\boldsymbol{\zeta}^s|)}{|\boldsymbol{\zeta}^s|^2}\right\}\frac{\zeta_\mu^s\zeta_\nu^s-\delta_{\mu\nu}|\boldsymbol{\zeta}^s|^2}{|\boldsymbol{\zeta}^s|^3}\right],$$

$$\text{(A.24)}$$

and in the case of $\mu = \nu = z$,

$$V_{H,\mu\nu}^{3,s}(\boldsymbol{r}) = \frac{\pi}{S} \sum_{|\boldsymbol{G}|\neq 0} \cos\left[\boldsymbol{G}\cdot(\boldsymbol{r}-\boldsymbol{R}^s)\right]$$

$$\times\left[|\boldsymbol{G}|\,f^+(\boldsymbol{G},\boldsymbol{r})-\frac{4\eta}{\sqrt{\pi}}\exp\left\{-\frac{|\boldsymbol{G}|^2+4(z-R_z^s)^2\eta^4}{4\eta^2}\right\}\right]$$

$$-\frac{4\sqrt{\pi}\eta}{S}\exp(-|z-R_z^s|^2\eta^2)$$

$$+ \sum_{P}\left[\left\{\frac{4\eta\exp(-\eta^2|\boldsymbol{\zeta}^s|^2)}{\sqrt{\pi}}\left(\eta^2+|\boldsymbol{\zeta}^s|^{-2}\right)+\frac{2\,\mathrm{erfc}(\eta|\boldsymbol{\zeta}^s|)}{|\boldsymbol{\zeta}^s|^3}\right\}\frac{\zeta_\mu^s\zeta_\nu^s}{|\boldsymbol{\zeta}^s|^2}\right.$$

$$\left.+\left\{\frac{2\eta\exp(-\eta^2|\boldsymbol{\zeta}^s|^2)}{\sqrt{\pi}|\boldsymbol{\zeta}^s|}+\frac{\mathrm{erfc}(\eta|\boldsymbol{\zeta}^s|)}{|\boldsymbol{\zeta}^s|^2}\right\}\frac{\zeta_\mu^s\zeta_\nu^s-\delta_{\mu\nu}|\boldsymbol{\zeta}^s|^2}{|\boldsymbol{\zeta}^s|^3}\right].$$

$$\text{(A.25)}$$

Coulomb energy among the nuclei

$$\gamma_E = \frac{\pi}{2S}\sum_{s,s'}Z_sZ_{s'}\sum_{|\boldsymbol{G}|\neq 0}\frac{1}{|\boldsymbol{G}|}\cos\left[\boldsymbol{G}\cdot(\boldsymbol{R}^{s'}-\boldsymbol{R}^s)\right]f^+(\boldsymbol{G},\boldsymbol{R}^{s'})$$

$$-\frac{\sqrt{\pi}}{S}\sum_{s,s'}Z_sZ_{s'}$$

$$\times\left[\frac{1}{\eta}\exp(-|R_z^{s'}-R_z^s|^2\eta^2)+\sqrt{\pi}\,|R_z^{s'}-R_z^s|\,\mathrm{erf}(|R_z^{s'}-R_z^s|\eta)\right]$$

$$-\sum_{s,s'}Z_sZ_{s'}\delta_{ss'}\frac{\eta}{\sqrt{\pi}}+\frac{1}{2}\sum_{P,s,s'}{}'Z_sZ_{s'}\frac{\mathrm{erfc}(\eta|\boldsymbol{\xi}^{s,s'}|)}{|\boldsymbol{\xi}^{s,s'}|}, \qquad \text{(A.26)}$$

where $\boldsymbol{\xi}^{s,s'} = \boldsymbol{P} + \boldsymbol{R}^{s'} - \boldsymbol{R}^s$.

The differentiations of γ_E are as follows: in the case of $\mu = x, y$,

$$-\frac{\partial \gamma_E}{\partial R^s_\mu} = -\frac{\pi}{2S} \sum_{s'} Z_s Z_{s'} \sum_{|G| \neq 0} \frac{G_\mu}{|G|} \sin\left[\boldsymbol{G} \cdot (\boldsymbol{R}^{s'} - \boldsymbol{R}^s)\right] f^+(\boldsymbol{G}, \boldsymbol{R}^{s'})$$

$$- \sum_{\boldsymbol{P}, s'}' Z_s Z_{s'} \left[\frac{\eta \exp(-\eta^2 |\boldsymbol{\xi}^{s,s'}|^2)}{\sqrt{\pi}\,|\boldsymbol{\xi}^{s,s'}|} + \frac{\mathrm{erfc}(\eta\,|\boldsymbol{\xi}^{s,s'}|)}{2|\boldsymbol{\xi}^{s,s'}|^2}\right] \frac{\xi^{s,s'}_\mu}{|\boldsymbol{\xi}^{s,s'}|}, \quad \text{(A.27)}$$

and in the case of $\mu = z$,

$$-\frac{\partial \gamma_E}{\partial R^s_\mu} = -\frac{\pi}{2S} \sum_{s'} Z_s Z_{s'} \sum_{|G| \neq 0} \cos\left[\boldsymbol{G} \cdot (\boldsymbol{R}^{s'} - \boldsymbol{R}^s)\right] f^-(\boldsymbol{G}, \boldsymbol{R}^{s'})$$

$$- \frac{\pi}{S} \sum_{s'} Z_s Z_{s'} \mathrm{erf}(|R^{s'}_z - R^s_z|\eta) \frac{(R^{s'}_z - R^s_z)}{|R^{s'}_z - R^s_z|}$$

$$- \sum_{\boldsymbol{P}, s'}' Z_s Z_{s'} \left[\frac{\eta \exp(-\eta^2 |\boldsymbol{\xi}^{s,s'}|^2)}{\sqrt{\pi}\,|\boldsymbol{\xi}^{s,s'}|} + \frac{\mathrm{erfc}(\eta\,|\boldsymbol{\xi}^{s,s'}|)}{2|\boldsymbol{\xi}^{s,s'}|^2}\right] \frac{\xi^{s,s'}_z}{|\boldsymbol{\xi}^{s,s'}|}. \quad \text{(A.28)}$$

A nucleus does not interact with its own Coulomb charge, so that the $\boldsymbol{P} = 0$ term must be omitted from the real-space summation when $s = s'$. The prime in the last summation in (A.26), (A.27), and (A.28) means that $|\boldsymbol{\xi}^{s,s'}| = 0$ is omitted.

A.3 (1D Periodic + 2D Isolated) Boundary Condition

The potential and energy in the case of a (1D periodic + 2D isolated) boundary condition are also obtained by the Ewald summation. Here, the case in which isolated boundary conditions are imposed in the x and y directions and a periodic boundary condition in the z direction is explained.

Ionic pseudopotential

By using the identity $(b > 0)$,

$$\int_0^1 \frac{1}{t} \exp\left(-t^2 |\boldsymbol{r}_{xy}|^2 - \frac{b^2}{4t^2}\right) dt$$

$$= \frac{1}{\pi} \exp\left(-\frac{b^2}{4}\right) \int_{-\infty}^{\infty} K_0\left(b \cdot |\boldsymbol{r}_{xy} - \boldsymbol{r}'_{xy}|^2\right) \exp\left(-|\boldsymbol{r}'_{xy}|^2\right) d\boldsymbol{r}'_{xy}, \quad \text{(A.29)}$$

the ionic pseudopotential is given by

$$V_{loc}^{s}(r) = -\frac{2\eta^2}{\pi L_z} Z_s \sum_{|\mathbf{G}|\neq 0} \cos\left[\mathbf{G} \cdot (\mathbf{r} - \mathbf{R}^s)\right] \exp\left(-\frac{|\mathbf{G}|^2}{4\eta^2}\right)$$

$$\times \int_{-\infty}^{\infty} K_0(|\mathbf{G}| \cdot |\mathbf{r}_{xy} - \mathbf{R}_{xy}^s - \mathbf{r}_{xy}'|) \exp(-\eta^2 |\mathbf{r}_{xy}'|^2)\, d\mathbf{r}_{xy}'$$

$$-\frac{1}{L_z} Z_s \left[\text{Ei}(-\eta^2 |\mathbf{r}_{xy} - \mathbf{R}_{xy}^s|^2) - \ln(|\mathbf{r}_{xy} - \mathbf{R}_{xy}^s|^2)\right]$$

$$+Z_s \sum_{\mathbf{P}} \frac{1}{|\boldsymbol{\zeta}^s|} \left[\text{erf}(\eta\,|\boldsymbol{\zeta}^s|) - \sum_{i=1,2} C_{s,i}\,\text{erf}(\sqrt{\alpha_{s,i}}\,|\boldsymbol{\zeta}^s|)\right],$$

$$\text{(A.30)}$$

where $\mathbf{G} = 2\pi(0, 0, \frac{j_z}{L_z})$, $\mathbf{P} = (0, 0, n_z L_z)$, $\mathbf{r}_{xy} - \mathbf{R}_{xy}^s = (x - R_x^s, y - R_y^s)$, $\boldsymbol{\zeta}^s = \mathbf{P} + \mathbf{r} - \mathbf{R}^s$, K_0 is the modified Bessel function, and Ei is the exponential-integral function.

Hartree potential

$V_H^{1,s}(r)$ is as follow:

$$V_H^{1,s}(r) = \frac{2\eta^2}{\pi L_z} \sum_{|\mathbf{G}|\neq 0} \cos\left[\mathbf{G} \cdot (\mathbf{r} - \mathbf{R}^s)\right] \exp\left(-\frac{|\mathbf{G}|^2}{4\eta^2}\right)$$

$$\times \int_{-\infty}^{\infty} K_0(|\mathbf{G}| \cdot |\mathbf{r}_{xy} - \mathbf{R}_{xy}^s - \mathbf{r}_{xy}'|) \exp(-\eta^2 |\mathbf{r}_{xy}'|^2)\, d\mathbf{r}_{xy}'$$

$$+\frac{1}{L_z} \left[\text{Ei}(-\eta^2 |\mathbf{r}_{xy} - \mathbf{R}_{xy}^s|^2) - \ln(|\mathbf{r}_{xy} - \mathbf{R}_{xy}^s|^2)\right]$$

$$+\sum_{\mathbf{P}} \frac{1}{|\boldsymbol{\zeta}^s|} \text{erfc}(\eta\,|\boldsymbol{\zeta}^s|). \qquad \text{(A.31)}$$

$V_{H,\mu}^{2,s}(r)$ are as follows: in the case of $\mu = z$,

$$V_{H,\mu}^{2,s}(r) = \frac{2\eta^2}{\pi L_z} \sum_{|\mathbf{G}|\neq 0} G_\mu \sin\left[\mathbf{G} \cdot (\mathbf{r} - \mathbf{R}^s)\right] \exp\left(-\frac{|\mathbf{G}|^2}{4\eta^2}\right)$$

$$\times \int_{-\infty}^{\infty} K_0(|\mathbf{G}| \cdot |\mathbf{r}_{xy} - \mathbf{R}_{xy}^s - \mathbf{r}_{xy}'|) \exp(-\eta^2 |\mathbf{r}_{xy}'|^2)\, d\mathbf{r}_{xy}'$$

$$+\sum_{\mathbf{P}} \left[\frac{2\eta \exp(-\eta^2 |\boldsymbol{\zeta}^s|^2)}{\sqrt{\pi}\,|\boldsymbol{\zeta}^s|} + \frac{\text{erfc}(\eta\,|\boldsymbol{\zeta}^s|)}{|\boldsymbol{\zeta}^s|^2}\right] \frac{\zeta_\mu^s}{|\boldsymbol{\zeta}^s|}, \qquad \text{(A.32)}$$

and in the case of $\mu = x, y$,

$$
\begin{aligned}
V_{H,\mu}^{2,s}(\boldsymbol{r}) = & \frac{2\eta^2}{\pi L_z} \sum_{|\boldsymbol{G}| \neq 0} \cos\left[\boldsymbol{G} \cdot (\boldsymbol{r} - \boldsymbol{R}^s)\right] \exp\left(-\frac{|\boldsymbol{G}|^2}{4\eta^2}\right) \\
& \times \int_{-\infty}^{\infty} \frac{\partial}{\partial R_\mu^s} K_0(|\boldsymbol{G}| \cdot |\boldsymbol{r}_{xy} - \boldsymbol{R}_{xy}^s - \boldsymbol{r}_{xy}'|) \exp(-\eta^2 |\boldsymbol{r}_{xy}'|^2) \, dr_{xy}' \\
& + \frac{2}{L_z} (\mu - R_\mu^s) \left[\frac{1 - \exp(-\eta^2 |\boldsymbol{r}_{xy} - \boldsymbol{R}_{xy}^s|^2)}{|\boldsymbol{r}_{xy} - \boldsymbol{R}_{xy}^s|^2} \right] \\
& + \sum_P \left[\frac{2\eta \exp(-\eta^2 |\boldsymbol{\zeta}^s|^2)}{\sqrt{\pi} |\boldsymbol{\zeta}^s|} + \frac{\operatorname{erfc}(\eta |\boldsymbol{\zeta}^s|)}{|\boldsymbol{\zeta}^s|^2} \right] \frac{\zeta_\mu^s}{|\boldsymbol{\zeta}^s|}.
\end{aligned} \tag{A.33}
$$

$V_{H,\mu\nu}^{3,s}(\boldsymbol{r})$ are as follows: in the case of $\mu = \nu = z$,

$$
\begin{aligned}
V_{H,\mu\nu}^{3,s}(\boldsymbol{r}) = & -\frac{2\eta^2}{\pi L_z} \sum_{|\boldsymbol{G}| \neq 0} G_\mu^2 \cos\left[\boldsymbol{G} \cdot (\boldsymbol{r} - \boldsymbol{R}^s)\right] \exp\left(-\frac{|\boldsymbol{G}|^2}{4\eta^2}\right) \\
& \times \int_{-\infty}^{\infty} K_0(|\boldsymbol{G}| \cdot |\boldsymbol{r}_{xy} - \boldsymbol{R}_{xy}^s - \boldsymbol{r}_{xy}'|) \exp(-\eta^2 |\boldsymbol{r}_{xy}'|^2) \, dr_{xy}' \\
& + \sum_P \left[\left\{ \frac{4\eta \exp(-\eta^2 |\boldsymbol{\zeta}^s|^2)}{\sqrt{\pi}} (\eta^2 + |\boldsymbol{\zeta}^s|^{-2}) + \frac{2 \operatorname{erfc}(\eta |\boldsymbol{\zeta}^s|)}{|\boldsymbol{\zeta}^s|^3} \right\} \frac{\zeta_\mu^s \zeta_\nu^s}{|\boldsymbol{\zeta}^s|^2} \right. \\
& \left. + \left\{ \frac{2\eta \exp(-\eta^2 |\boldsymbol{\zeta}^s|^2)}{\sqrt{\pi} |\boldsymbol{\zeta}^s|} + \frac{\operatorname{erfc}(\eta |\boldsymbol{\zeta}^s|)}{|\boldsymbol{\zeta}^s|^2} \right\} \frac{\zeta_\mu^s \zeta_\nu^s - \delta_{\mu\nu} |\boldsymbol{\zeta}^s|^2}{|\boldsymbol{\zeta}^s|^3} \right],
\end{aligned} \tag{A.34}
$$

in the case of $\mu = z$ and $\nu = x, y$,

$$
\begin{aligned}
V_{H,\mu\nu}^{3,s}(\boldsymbol{r}) = & \frac{2\eta^2}{\pi L_z} \sum_{|\boldsymbol{G}| \neq 0} G_\mu \sin\left[\boldsymbol{G} \cdot (\boldsymbol{r} - \boldsymbol{R}^s)\right] \exp\left(-\frac{|\boldsymbol{G}|^2}{4\eta^2}\right) \\
& \times \int_{-\infty}^{\infty} \frac{\partial}{\partial R_\nu^s} K_0(|\boldsymbol{G}| \cdot |\boldsymbol{r}_{xy} - \boldsymbol{R}_{xy}^s - \boldsymbol{r}_{xy}'|) \exp(-\eta^2 |\boldsymbol{r}_{xy}'|^2) \, dr_{xy}' \\
& + \sum_P \left[\left\{ \frac{4\eta \exp(-\eta^2 |\boldsymbol{\zeta}^s|^2)}{\sqrt{\pi}} (\eta^2 + |\boldsymbol{\zeta}^s|^{-2}) + \frac{2 \operatorname{erfc}(\eta |\boldsymbol{\zeta}^s|)}{|\boldsymbol{\zeta}^s|^3} \right\} \frac{\zeta_\mu^s \zeta_\nu^s}{|\boldsymbol{\zeta}^s|^2} \right. \\
& \left. + \left\{ \frac{2\eta \exp(-\eta^2 |\boldsymbol{\zeta}^s|^2)}{\sqrt{\pi} |\boldsymbol{\zeta}^s|} + \frac{\operatorname{erfc}(\eta |\boldsymbol{\zeta}^s|)}{|\boldsymbol{\zeta}^s|^2} \right\} \frac{\zeta_\mu^s \zeta_\nu^s - \delta_{\mu\nu} |\boldsymbol{\zeta}^s|^2}{|\boldsymbol{\zeta}^s|^3} \right],
\end{aligned} \tag{A.35}
$$

and in the case of $\mu = x, y$ and $\nu = x, y$,

$$V^{3,s}_{H,\mu\nu}(\boldsymbol{r})$$

$$= \frac{2\eta^2}{\pi L_z} \sum_{|\boldsymbol{G}|\neq 0} \cos\left[\boldsymbol{G} \cdot (\boldsymbol{r} - \boldsymbol{R}^s)\right] \exp\left(-\frac{|\boldsymbol{G}|^2}{4\eta^2}\right)$$

$$\times \int_{-\infty}^{\infty} \frac{\partial^2}{\partial R^s_\mu \partial R^s_\nu} K_0(|\boldsymbol{G}| \cdot |\boldsymbol{r}_{xy} - \boldsymbol{R}^s_{xy} - \boldsymbol{r}'_{xy}|) \exp(-\eta^2|\boldsymbol{r}'_{xy}|^2)\, d\boldsymbol{r}'_{xy}$$

$$- \frac{2\delta_{\mu\nu}}{L_z}\left[\frac{1 - \exp(-\eta^2|\boldsymbol{r}_{xy} - \boldsymbol{R}^s_{xy}|^2)}{|\boldsymbol{r}_{xy} - \boldsymbol{R}^s_{xy}|^2}\right] + \frac{4}{L_z}(\mu - R^s_\mu)(\nu - R^s_\nu)$$

$$\times \left[\frac{1 - (1 + \eta^2|\boldsymbol{r}_{xy} - \boldsymbol{R}^s_{xy}|^2)\exp(-\eta^2|\boldsymbol{r}_{xy} - \boldsymbol{R}^s_{xy}|^2)}{|\boldsymbol{r}_{xy} - \boldsymbol{R}^s_{xy}|^4}\right]$$

$$+ \sum_{\boldsymbol{P}}\left[\left\{\frac{4\eta \exp(-\eta^2|\boldsymbol{\zeta}^s|^2)}{\sqrt{\pi}}\left(\eta^2 + |\boldsymbol{\zeta}^s|^{-2}\right) + \frac{2\,\mathrm{erfc}(\eta|\boldsymbol{\zeta}^s|)}{|\boldsymbol{\zeta}^s|^3}\right\} \frac{\zeta^s_\mu \zeta^s_\nu}{|\boldsymbol{\zeta}^s|^2}\right.$$

$$\left. + \left\{\frac{2\eta \exp(-\eta^2\,|\boldsymbol{\zeta}^s|^2)}{\sqrt{\pi}\,|\boldsymbol{\zeta}^s|} + \frac{\mathrm{erfc}(\eta\,|\boldsymbol{\zeta}^s|)}{|\boldsymbol{\zeta}^s|^2}\right\} \frac{\zeta^s_\mu \zeta^s_\nu - \delta_{\mu\nu}|\boldsymbol{\zeta}^s|^2}{|\boldsymbol{\zeta}^s|^3}\right]. \tag{A.36}$$

Coulomb energy among the nuclei

γ_E is as follow:

$$\gamma_E = \frac{\eta^2}{\pi L_z} \sum_{s,s'} Z_s Z_{s'} \sum_{|\boldsymbol{G}|\neq 0} \cos\left[\boldsymbol{G} \cdot (\boldsymbol{R}^{s'} - \boldsymbol{R}^s)\right] \exp\left(-\frac{|\boldsymbol{G}|^2}{4\eta^2}\right)$$

$$\times \int_{-\infty}^{\infty} K_0(|\boldsymbol{G}| \cdot |\boldsymbol{R}^{s'}_{xy} - \boldsymbol{R}^s_{xy} - \boldsymbol{r}'_{xy}|) \exp(-\eta^2|\boldsymbol{r}'_{xy}|^2)\, d\boldsymbol{r}'_{xy}$$

$$+ \frac{1}{2L_z} \sum_{s,s'} Z_s Z_{s'} \left[\mathrm{Ei}(-\eta^2|\boldsymbol{R}^{s'}_{xy} - \boldsymbol{R}^s_{xy}|^2) - \ln(|\boldsymbol{R}^{s'}_{xy} - \boldsymbol{R}^s_{xy}|^2)\right]$$

$$- \sum_{s,s'} Z_s Z_{s'} \delta_{ss'} \frac{\eta}{\sqrt{\pi}} + \frac{1}{2} \sum_{\boldsymbol{P},s,s'}' Z_s Z_{s'} \frac{\mathrm{erfc}(\eta\,|\boldsymbol{\xi}^{s,s'}|)}{|\boldsymbol{\xi}^{s,s'}|}, \tag{A.37}$$

where $\boldsymbol{\xi}^{s,s'} = \boldsymbol{P} + \boldsymbol{R}^{s'} - \boldsymbol{R}^s$.

The differentiations of γ_E are as follows: in the case of $\mu = z$,

$$-\frac{\partial \gamma_E}{\partial R_\mu^s} = -\frac{\eta^2}{\pi L_z} \sum_{s'} Z_s Z_{s'} \sum_{|\boldsymbol{G}| \neq 0} G_\mu \sin\left[\boldsymbol{G} \cdot (\boldsymbol{R}^{s'} - \boldsymbol{R}^s)\right] \exp\left(-\frac{|\boldsymbol{G}|^2}{4\eta^2}\right)$$

$$\times \int_{-\infty}^{\infty} K_0(|\boldsymbol{G}| \cdot |\boldsymbol{R}_{xy}^{s'} - \boldsymbol{R}_{xy}^s - \boldsymbol{r}_{xy}'|) \exp(-\eta^2 |\boldsymbol{r}_{xy}'|^2) \, dr_{xy}'$$

$$- \sum_{\boldsymbol{P}, s'}' Z_s Z_{s'} \left[\frac{\eta \exp(-\eta^2 |\boldsymbol{\xi}^{s,s'}|^2)}{\sqrt{\pi} \, |\boldsymbol{\xi}^{s,s'}|} + \frac{\text{erfc}(\eta \, |\boldsymbol{\xi}^{s,s'}|)}{2|\boldsymbol{\xi}^{s,s'}|^2}\right] \frac{\xi_\mu^{s,s'}}{|\boldsymbol{\xi}^{s,s'}|},$$

$$(A.38)$$

and in the case of $\mu = x, y$,

$$-\frac{\partial \gamma_E}{\partial R_\mu^s} = -\frac{\eta^2}{\pi L_z} \sum_{s'} Z_s Z_{s'} \sum_{|\boldsymbol{G}| \neq 0} \cos\left[\boldsymbol{G} \cdot (\boldsymbol{R}^{s'} - \boldsymbol{R}^s)\right] \exp\left(-\frac{|\boldsymbol{G}|^2}{4\eta^2}\right)$$

$$\times \int_{-\infty}^{\infty} \frac{\partial}{\partial R_\mu^s} K_0(|\boldsymbol{G}| \cdot |\boldsymbol{R}_{xy}^{s'} - \boldsymbol{R}_{xy}^s - \boldsymbol{r}_{xy}'|) \exp(-\eta^2 |\boldsymbol{r}_{xy}'|^2) \, dr_{xy}'$$

$$- \frac{1}{L_z} \sum_{s'} Z_s Z_{s'} (R_\mu^{s'} - R_\mu^s) \left[\frac{1 - \exp(-\eta^2 |\boldsymbol{R}_{xy}^{s'} - \boldsymbol{R}_{xy}^s|^2)}{|\boldsymbol{R}_{xy}^{s'} - \boldsymbol{R}_{xy}^s|^2}\right]$$

$$- \sum_{\boldsymbol{P}, s'}' Z_s Z_{s'} \left[\frac{\eta \exp(-\eta^2 |\boldsymbol{\xi}^{s,s'}|^2)}{\sqrt{\pi} \, |\boldsymbol{\xi}^{s,s'}|} + \frac{\text{erfc}(\eta \, |\boldsymbol{\xi}^{s,s'}|)}{2|\boldsymbol{\xi}^{s,s'}|^2}\right] \frac{\xi_\mu^{s,s'}}{|\boldsymbol{\xi}^{s,s'}|}.$$

$$(A.39)$$

A nucleus does not interact with its own Coulomb charge, so that the $\boldsymbol{P} = 0$ term must be omitted from the real-space summation when $s = s'$. The prime in the last summation in (A.37), (A.38), and (A.39) means that $|\boldsymbol{\xi}^{s,s'}| = 0$ is omitted.

A.4   Twist Boundary Condition

Let us consider the helically periodic model shown in Fig. A.1, in which a translation L_z units up the z axis is conjunct with a right-handed rotation φ about the z axis, where L_z is the length of the irreducible unit cell. We define the integral m_z so that a translation $m_z L_z$ becomes pure translation, i.e., $\varphi = 0$. The ionic pseudopotential, which is described by a similar

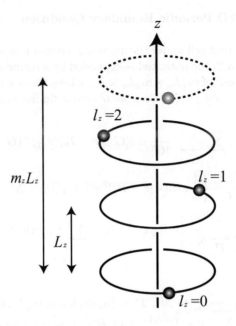

Fig. A.1   Schematic image of twist boundary condition.

equation to (A.30), is

$$V_{loc}^s(\boldsymbol{r}) = -\frac{2\eta^2}{\pi m_z L_z} Z_s \sum_{l_z=0}^{m_z-1} \sum_{|\boldsymbol{G}|\neq 0} \cos\left[\boldsymbol{G}\cdot(\boldsymbol{r}-\boldsymbol{R}^{s,l_z})\right] \exp(-\frac{|\boldsymbol{G}|^2}{4\eta^2})$$

$$\times \int_{-\infty}^{\infty} K_0(|\boldsymbol{G}|\cdot|\boldsymbol{r}_{xy}-\boldsymbol{R}_{xy}^{s,l_z}-\boldsymbol{r}'_{xy}|)\exp(-\eta^2|\boldsymbol{r}'_{xy}|)\,d\boldsymbol{r}'_{xy}$$

$$-\frac{1}{m_z L_z} Z_s \sum_{l_z=0}^{m_z-1}\left[\mathrm{Ei}(-\eta^2\,|\boldsymbol{r}_{xy}-\boldsymbol{R}_{xy}^{s,l_z}|^2)-\ln(|\boldsymbol{r}_{yz}-\boldsymbol{R}_{yz}^{s,l_z}|^2)\right]$$

$$+Z_s\sum_{l_z=0}^{m_z-1}\sum_{\boldsymbol{P}}\frac{1}{|\boldsymbol{\zeta}^{s,l_z}|}\left[\mathrm{erf}(\eta\,|\boldsymbol{\zeta}^s|)-\sum_{i=1,2}C_{s,i}\,\mathrm{erf}(\sqrt{\alpha_{s,i}}\,|\boldsymbol{\zeta}^s|)\right],$$

$$(A.40)$$

where $\boldsymbol{G} = 2\pi(0,0,\frac{j_z}{m_z L_z})$, $\boldsymbol{P} = (0,0,n_z m_z L_z)$, $\boldsymbol{r}_{xy}-\boldsymbol{R}_{xy}^s = (x-R_x^s, y-R_y^s)$, $\boldsymbol{\zeta}^{s,l_z} = \boldsymbol{P}+\boldsymbol{r}-\boldsymbol{R}^{s,l_z}$, and $\boldsymbol{R}^{s,l_z}$ is the atomic coordinate of the s-th atom in the l_z-th irreducible unit cell.

A.5 Uneven 2D Periodic Boundary Condition

In the case of the unit cell under the uneven 2D periodic boundary condition as shown in Fig. A.2, the potential is expanded by a reciprocal lattice-vector in the large unit cell of $m_x L_x \times m_y L_y \times L_z$, where the size of the irreducible unit cell size is $L_x \times L_y \times L_z$, then we obtain a similar equation to (A.13)

$$
V_{loc}^s(\boldsymbol{r}) = -\frac{\pi}{S} Z_s \sum_{l_x,l_y} \sum_{|\boldsymbol{G}|\neq 0} \frac{1}{|\boldsymbol{G}|} \cos[\boldsymbol{G} \cdot (\boldsymbol{r} - \boldsymbol{R}_s^{l_x l_y})] f^+(\boldsymbol{G}, \boldsymbol{r})
$$
$$
+ \frac{2\sqrt{\pi} Z_s}{L_x L_y} \left\{ \frac{1}{\eta} \exp(-|z - R_z^s|^2 \eta^2) + \sqrt{\pi}|z - R_z^s| \mathrm{erf}(|z - R_z^s|\eta) \right\}
$$
$$
+ \sum_{l_x,l_y} \sum_{\boldsymbol{P}} \frac{Z_s}{|\boldsymbol{\zeta}^s|} \left[\mathrm{erf}(\eta|\boldsymbol{\zeta}^{s,l_x l_y}|) - \sum_{i=1,2} C_{s,i} \mathrm{erf}(\sqrt{\alpha_{s,i}}|\boldsymbol{\zeta}^{s,l_x l_y}|) \right],
$$
$$
\tag{A.41}
$$

where $\boldsymbol{G} = 2\pi(\frac{j_x}{m_x L_x}, \frac{j_y}{m_y L_y}, 0)$, $\boldsymbol{P} = (n_x m_x L_x, n_y m_y L_y, 0)$, $S = m_x L_x \times m_y L_y$, $\boldsymbol{\zeta}^{s,l_x l_y} = \boldsymbol{P} + \boldsymbol{r} - \boldsymbol{R}^{s,l_x l_y}$, and $\boldsymbol{R}^{s,l_x l_y}$ is the atomic coordinate of the s-th atom of the l_x-th and l_y-th irreducible unit cell in the x and y

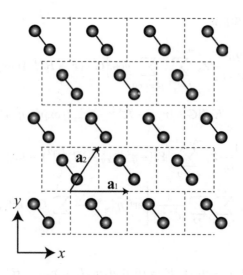

Fig. A.2 Schematic image of uneven 2D periodic boundary condition. The x axis is taken to be parallel to the $\mathbf{a}_1$ vector.

directions, respectively. Since

$$\sum_{l_x=0}^{m_x-1} \sum_{l_y=0}^{m_y-1} \cos \left[\boldsymbol{G} \cdot (\boldsymbol{r} - \boldsymbol{R}_s^{l_x l_y}) \right]$$

$$= \sum_{l_x=0}^{m_x-1} \sum_{l_y=0}^{m_y-1} \cos 2\pi \left(j_x \frac{l_x L_x + l_y \frac{m_x L_x}{m_y} + x - R_x^s}{m_x L_x} + j_y \frac{l_y L_y + y - R_y^s}{m_y L_y} \right)$$

$$= \sum_{l_x=0}^{m_x-1} \sum_{l_y=0}^{m_y-1} \cos 2\pi \left[\left\{ j_x \frac{l_x}{m_x} + (j_x + j_y) \frac{l_y}{m_y} \right\} \right.$$

$$\left. + \left(j_x \frac{x - R_x^s}{m_x L_x} + j_y \frac{y - R_y^s}{m_y L_y} \right) \right],$$

$$(A.42)$$

the summation regarding l_x and l_y in the first term of (A.41) is implemented when both j_x/m_x and $(j_x + j_y)/m_y$ are integers.

A.6 Uneven 3D Periodic Boundary Condition

As well as the case of the uneven 2D periodic boundary condition, the ionic pseudopotential is given by a similar equation to (A.6)

$$V_{loc}^s(\boldsymbol{r}) = -\frac{4\pi}{\Omega} Z_s \sum_{l_x, l_y, l_z} \sum_{\boldsymbol{G} \neq 0} \frac{1}{|\boldsymbol{G}|^2} \cos \left[\boldsymbol{G} \cdot (\boldsymbol{r} - \boldsymbol{R}^{s, l_x l_y l_z}) \right] \exp \left(-\frac{|\boldsymbol{G}|^2}{4\eta^2} \right)$$

$$+ \frac{\pi Z_s}{L_x L_y L_z \eta^2} + \sum_{l_x, l_y, l_z} \sum_{\boldsymbol{P}} \frac{Z_s}{|\boldsymbol{\zeta}^{s, l_x l_y l_z}|}$$

$$\times \left[\text{erf}(\eta \, |\boldsymbol{\zeta}^{s, l_x l_y l_z}|) - \sum_{i=1,2} C_{s,i} \, \text{erf}(\sqrt{\alpha_{s,i}} \, |\boldsymbol{\zeta}^{s, l_x l_y l_z}|) \right], \qquad (A.43)$$

where $\boldsymbol{G} = 2\pi(\frac{j_x}{m_x L_x}, \frac{j_y}{m_y L_y}, \frac{j_z}{m_z L_z})$, $\boldsymbol{P} = (n_x m_x L_x, n_y m_y L_y, n_z m_z L_z)$, $\Omega = m_x L_x \times m_y L_y \times m_z L_z$, $\boldsymbol{\zeta}^{s, l_x l_y l_z} = \boldsymbol{P} + \boldsymbol{r} - \boldsymbol{R}^{s, l_x l_y l_z}$, and $\boldsymbol{R}^{s, l_x l_y l_z}$ is the atomic coordinate of the s-th atom of the l_x-th, l_y-th and l_z-th irreducible unit cell in the x, y, and z directions, respectively. Here, we chose the x axis parallel to the $\boldsymbol{a}_1$ vector and the z axis normal to the plane consisting of the $\boldsymbol{a}_1$ and $\boldsymbol{a}_2$ vectors. The summation regarding l_x, l_y and l_z in the first term of (A.43) is implemented when j_x/m_x, $(j_x + j_y)/m_y$, and $(j_x + j_z)/m_z$ are integers.

Appendix B

Tight-Binding Approach Based on the Overbridging Boundary-Matching Scheme

Kikuji Hirose

The overbridging boundary-matching (OBM) procedure developed in Part II, which is an *ab initio* procedure for calculating the electronic configurations and transport properties of nanostructures attached to semi-infinite electrodes (Fig. B.1), can also be formulated within the tight-binding (TB) approach. We here outline this procedure. (For brief discussions in Sections B.1, B.2 and B.3, readers should refer to more detailed arguments in Sections 6.3 – 6.5, 6.2 and 9.3 – 9.5, respectively, with some apparent modifications.) The TB approach is efficient to investigate transport properties of large systems using the nonequilibrium Green's function (NEGF) formula [see Eq. (B.29)]. In addition, (B.27) and (B.28) derived below from

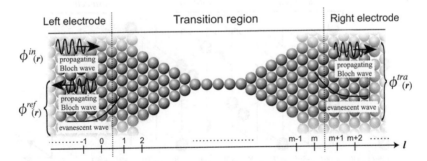

Fig. B.1 Sketch of a system with the transition region sandwiched between the left and right semi-infinite crystalline electrodes. The indices l are for principal layers (see Section B.2). The case of the incident wave ϕ^{in} coming from the left electrode is illustrated. In the left (right) electrode, ϕ^{ref} (ϕ^{tra}) denotes the reflected (transmitted) waves consisting of the propagating and decaying evanescent waves toward the left (right) side. Two-dimensional periodicity in the directions perpendicular to the nanoscale junction is assumed.

the OBM consideration are of great use for treating still larger systems, and the wave-function matching formula (B.21) and the related formulas (B.32) – (B.35) are further useful for determining scattering wave functions. The calculation accuracy in the TB approach is improved by adopting a basis set of quality; however, precise descriptions of electronic states are difficult, particularly in tunneling regions. The real-space finite-difference approach is advantageous to the analysis of phenomena in which tunneling effects dominate, such as leak currents through oxidized membranes, tunneling currents in scanning tunneling microscopies, and resonant tunneling effects in quantum-dot structures.

B.1   Generalized Bloch States inside Semi-Infinite Crystalline Electrodes

It is known that any three-dimensional crystal can be viewed as an infinite stack of principal layers [Lee and Joannopoulos (1981b)]. A principal layer is defined as the *smallest* group of atomic layers such that only nearest-neighbor interactions between principal layers exist and principal layers form a periodic array of unit cells of periodicity. A simple example is shown in Fig. B.2 for a diamond-structured bulk; when nearest- (second-)neighbor interactions are taken into account between 'layer orbitals', a principal layer

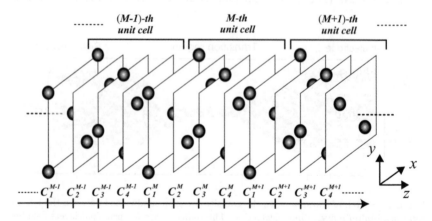

Fig. B.2   Schematic view of a crystal regarded as an infinite stack of principal layers. The case of a diamond-structured bulk with nearest-neighbor interactions between atomic orbitals is shown. C_l^M represents the vector consisting of the expansion coefficients relative to the l-th principal layer in the M-th unit cell, where $l = 1, 2, ..., m$ ($m = 4$ for the present case).

consists of one (two) atomic layer and four (two) principal layers constitute one unit cell. We assume that the numbers of inequivalent atomic orbitals in the respective principal layers are *equally* N, for simplicity, and that there are m principal layers in each unit cell. In addition, we note that two-dimensional periodicity parallel to the layers reduces the original system to a noninteracting linear chain of principal layers every lateral Bloch wave vector $\mathbf{k}_{||}$.

Within a linear combination of atomic orbitals (LCAO) framework, any wave function for the linear chain of principal layers, $\phi(\mathbf{r}; E, \mathbf{k}_{||})$, may be expanded as

$$\phi(\mathbf{r}; E, \mathbf{k}_{||}) = \sum_{M=-\infty}^{\infty} \sum_{l=1}^{m} \sum_{\mu=1}^{N} c_{l,\mu}^{M}(E, \mathbf{k}_{||}) \, \alpha_{l,\mu}^{M}(\mathbf{r}; E, \mathbf{k}_{||}) \tag{B.1}$$

in terms of the Bloch basis functions

$$\alpha_{l,\mu}^{M}(\mathbf{r}; E, \mathbf{k}_{||}) = \frac{1}{\sqrt{N}} \sum_{\mathbf{P}_{||}} e^{i\mathbf{k}_{||}(\mathbf{P}_{||}+\mathbf{R}_{l,\mu}^{M})} \chi_{l,\mu}^{M}(\mathbf{r} - \mathbf{P}_{||} - \mathbf{R}_{l,\mu}^{M}), \tag{B.2}$$

where $\chi_{l,\mu}^{M}$ is the μ-th atomic orbital in the l-th principal layer in the M-th unit cell, $\mathbf{R}_{l,\mu}^{M}$ is the corresponding atomic position, and $\mathbf{P}_{||} = (P_x, P_y, 0)$ is a lateral lattice vector. Hereafter, we treat the set of expansion coefficients $\{c_{l,\mu}^{M}(E, \mathbf{k}_{||})\}$ as an N-dimensional columnar vector

$$C_{l}^{M}(E, \mathbf{k}_{||}) = \left[c_{l,1}^{M}(E, \mathbf{k}_{||}), c_{l,2}^{M}(E, \mathbf{k}_{||}), ..., c_{l,N}^{M}(E, \mathbf{k}_{||}) \right]^{t}. \tag{B.3}$$

The Kohn–Sham equation can then be written in the form of a three-term matrix equation with respect to $\{C_{l}^{M}\}$,

$$-B_{l-1}^{M}{}^{\dagger} C_{l-1}^{M} + (ES_{l}^{M} - H_{l}^{M})C_{l}^{M} - B_{l}^{M} C_{l+1}^{M} = 0, \tag{B.4}$$

where the TB matrix elements are defined by

$$(H_{l}^{M})_{\mu\nu} = \left\langle \alpha_{l,\mu}^{M} \middle| \hat{H} \middle| \alpha_{l,\nu}^{M} \right\rangle$$

$$(S_{l}^{M})_{\mu\nu} = \left\langle \alpha_{l,\mu}^{M} \middle| \alpha_{l,\nu}^{M} \right\rangle$$

$$(B_{l}^{M})_{\mu\nu} = \begin{cases} \left\langle \alpha_{l,\mu}^{M} \middle| \hat{H} \middle| \alpha_{l+1,\nu}^{M} \right\rangle - E\left\langle \alpha_{l,\mu}^{M} \middle| \alpha_{l+1,\nu}^{M} \right\rangle & \text{for } l = 1, 2, \cdots, m-1 \\[2mm] \left\langle \alpha_{m,\mu}^{M} \middle| \hat{H} \middle| \alpha_{1,\nu}^{M+1} \right\rangle - E\left\langle \alpha_{m,\mu}^{M} \middle| \alpha_{1,\nu}^{M+1} \right\rangle & \text{for } l = m. \end{cases}$$

$$\tag{B.5}$$

Here and hereafter, we omit the dependence of the coefficient vector and matrix elements on E and $k_{||}$ for the sake of clarity.

Equation (B.4) has exactly the same structure as that illustrated in Fig. 6.5 for the discretized Kohn–Sham equation in the real-space finite-difference approach. This allows us to proceed along the same line as the discussion in Section 6.3. Thus, on the basis of the OBM scheme, the Kohn–Sham equation (B.4) is transformed into the following $2N$-dimensional generalized eigenvalue equation [see Eq. (6.35) in Theorem 6.1] for determining the generalized Bloch states, i.e., the solutions of the Kohn–Sham equation inside semi-infinite electrodes in which the potentials are periodic:

$$\Pi_1(E, k_{||}) \begin{bmatrix} C_m^{M-1} \\ C_1^{M+1} \end{bmatrix} = \lambda \Pi_2(E, k_{||}) \begin{bmatrix} C_m^{M-1} \\ C_1^{M+1} \end{bmatrix} \tag{B.6}$$

with

$$\Pi_1(E, k_{||}) = \begin{bmatrix} \mathcal{W}_{m,1}^M & \mathcal{W}_{m,m}^M \\ 0 & I \end{bmatrix}$$

$$\Pi_2(E, k_{||}) = \begin{bmatrix} I & 0 \\ \mathcal{W}_{1,1}^M & \mathcal{W}_{1,m}^M \end{bmatrix}. \tag{B.7}$$

Here,

$$\begin{aligned} \mathcal{W}_{1,1}^M &= \mathcal{G}_T^M(1,1) B_m^{M\dagger} \\ \mathcal{W}_{1,m}^M &= \mathcal{G}_T^M(1,m) B_m^M \\ \mathcal{W}_{m,1}^M &= \mathcal{G}_T^M(m,1) B_m^{M\dagger} \\ \mathcal{W}_{m,m}^M &= \mathcal{G}_T^M(m,m) B_m^M, \end{aligned} \tag{B.8}$$

and $\mathcal{G}_T^M(k, l)$ is the N-dimensional (k, l) block-matrix element of the Green's function matrix,

$$\hat{\mathcal{G}}_T^M(E, k_{||}) = \left[E \hat{S}_T^M - \hat{H}_T^M \right]^{-1}, \tag{B.9}$$

associated with the k-th and l-th principal layers in the M-th unit cell, and $\hat{H}_T^M$ and $\hat{S}_T^M$ are the mN-dimensional submatrices of the truncated TB Hamiltonian and overlapping integrals restricted to the M-th unit cell, respectively.

The $2N$ eigenvectors of (B.6) are equally divided into two groups of N eigenvectors (see Theorem 6.6). One group consists of eigenvectors describing the states that propagate to the left if $|\lambda| = 1$ or decay to the left if $|\lambda| > 1$, while the other group of eigenvectors corresponds to the states propagating to the right if $|\lambda| = 1$ or decaying to the right if $|\lambda| < 1$. The former (latter) group is then the set of the solutions of the Kohn–Sham equation which are physical *inside* the left (right) semi-infinite electrode; thus, we denote this set as $\{C_{l,i}^{M,ref}\}$ ($\{C_{l,i}^{M,tra}\}$), where the superscript '*ref*' ('*tra*') stands for its correspondence to the reflected (transmitted) waves within the left (right) electrode in the case of the incident coming from the left electrode as illustrated in Fig. B.1, and $i = 1, 2, ..., N$, which is the numbering of the eigenvectors.

The advantage of the method employing the generalized eigenvalue equation (B.6) is that all of the generalized Bloch states at an assigned energy E and lateral Bloch wave vector $\mathbf{k}_{\parallel}$ are simultaneously obtainable. Equation (B.6) can be numerically solved by the QZ algorithm, a standard solver for a generalized eigenvalue problem. Up to now, there have been similar attempts to determine the generalized Bloch states based on recursive calculation techniques, which are called the transfer-matrix methods. These methods, however, may turn to error accumulation in the recursively multiplicated matrix elements as the number of basis functions increases, due to the appearance of exponentially growing and decaying evanescent waves (generalized Bloch states with $|\lambda| \neq 1$). On the contrary, (B.6) is always free from this numerical deficit and possesses well-defined matrix elements, since it is formulated by treating the Kohn–Sham equation as a boundary-value problem instead of an initial-value problem in the transfer-matrix methods. Nevertheless, in some cases, (B.6) gives rise to a numerical difficulty such that rapidly growing and decaying evanescent waves are smeared with numerical errors, even though propagating Bloch waves and gently growing or decaying evanescent waves are satisfactorily calculated.

An effective remedy for this numerical instability is to introduce the ratios of the generalized Bloch states at two successive principal layers. Let us define the N-dimensional ratio matrices

$$R_l^{M,ref} = Q_{l-1}^{M,ref}\left(Q_l^{M,ref}\right)^{-1} \quad \text{and} \quad R_l^{M,tra} = Q_l^{M,tra}\left(Q_{l-1}^{M,tra}\right)^{-1}, \tag{B.10}$$

where

$$Q_l^{M,A} = \left[C_{l,1}^{M,A}, C_{l,2}^{M,A}, ..., C_{l,N}^{M,A}\right] \quad (A = ref \text{ and } tra). \tag{B.11}$$

The ratio matrices obey the following continued-fraction equations in the OBM scheme and generalized Bloch conditions [see Eqs. (6.76) and (6.77) in Theorem 6.4]:

$$R_1^{M+1,ref} = \mathcal{W}_{m,m}^M + \mathcal{W}_{m,1}^M\left[\left(R_1^{M,ref}\right)^{-1} - \mathcal{W}_{1,1}^M\right]^{-1}\mathcal{W}_{1,m}^M$$

$$R_1^{M,tra} = \mathcal{W}_{1,1}^M + \mathcal{W}_{1,m}^M\left[\left(R_1^{M+1,tra}\right)^{-1} - \mathcal{W}_{m,m}^M\right]^{-1}\mathcal{W}_{m,1}^M, \tag{B.12}$$

$$R_l^{M+1,ref} = R_l^{M,ref} \quad \text{and} \quad R_l^{M+1,tra} = R_l^{M,tra}. \qquad \text{(B.13)}$$

A self-consistent calculation requiring that the ratio matrices should satisfy (B.12) under the restrictions of (B.13) yields the convergence of the ratio matrices to the correct solutions with several iterations, when the crude solutions of (B.6) are used as initial guesses of the self-consistent calculation. Hereafter, the ratio matrices are denoted simply by R_l^{ref} and R_l^{tra}, since they are independent of M. Thus, any numerical problems arising from evanescent waves with exponential behavior are completely excluded. The accurate ratio matrices thus determined are employed in the wave-function matching formula [see Eq. (B.21) in Section B.2] and in the Green's functions formalism [see Eq. (B.26) in Section B.3].

B.2   Wave-Function Matching Procedure

Similarly to a crystal in Fig. B.2, a whole system lacking the periodicity in the direction parallel to the nanoscale junction as indicated in Fig. B.1 is also regarded as an infinite stack of principal layers. By the ansatz of the LCAO formalism, a scattering wave function extending over the whole system, $\psi(\boldsymbol{r}; E, \boldsymbol{k}_{\|})$, is expanded as

$$\psi(\boldsymbol{r}; E, \boldsymbol{k}_{\|}) = \sum_{l=-\infty}^{\infty} \sum_{\mu=1}^{N_l} c_{l,\mu} \, \alpha_{l,\mu}(\boldsymbol{r}; E, \boldsymbol{k}_{\|}). \qquad \text{(B.14)}$$

Here, $\alpha_{l,\mu}$ is the Bloch basis function of the μ-th atomic orbital in the l-th principal layer [cf. Eq. (B.2)], and N_l is the number of inequivalent atomic orbitals in the l-th principal layer. The set of expansion coefficients $\{c_{l,\mu}\}$ is treated as an N_l-dimensional columnar vector

$$C_l = \left[c_{l,1}, c_{l,2}, ..., c_{l,N_l} \right]^t. \qquad \text{(B.15)}$$

We assume that N_l of the principal layer in the transition region is an arbitrary number in general, although N_l of any principal layer in the electrode regions is equally N as was postulated in Section B.1; therefore, the structure of the Hamiltonian of the whole system is the same as that illustrated in Fig. 9.2. It should be noted that the block tridiagonality is not needed for the TB Hamiltonian submatrix $\hat{H}_T$ and overlapping-integrals submatrix $\hat{S}_T$ in the transition region [precisely, those in the area from the 0-th principal layer to the $(m+1)$-th principal layer shown in Fig. B.1] to realize wave-function matching within the framework of the OBM scheme. Thus, the Kohn–Sham equation in the relevant area is written in the matrix form of

$$
\left[E\hat{S}_T - \hat{H}_T \right]
\begin{bmatrix}
C_0 \\
C_1 \\
\vdots \\
C_m \\
C_{m+1}
\end{bmatrix}
=
\begin{bmatrix}
B_{-1}^\dagger C_{-1} \\
0 \\
\vdots \\
0 \\
B_{m+1} C_{m+2}
\end{bmatrix},
\tag{B.16}
$$

which is a similar equation to (6.9). For the sake of brevity, explicit formulas for $\hat{H}_T$, $\hat{S}_T$, B_{-1} and B_{m+1} are omitted here. From (B.16), the relationship among the coefficient vectors near the surfaces of the electrodes, C_{-1}, C_0, C_{m+1} and C_{m+2}, is derived as

$$
\begin{bmatrix}
C_0 \\
C_{m+1}
\end{bmatrix}
=
\begin{bmatrix}
\mathcal{G}_T(0,0) & \mathcal{G}_T(0, m+1) \\
\mathcal{G}_T(m+1, 0) & \mathcal{G}_T(m+1, m+1)
\end{bmatrix}
\begin{bmatrix}
C_{-1} \\
C_{m+2}
\end{bmatrix},
\tag{B.17}
$$

where $\mathcal{G}_T(k, l)$ is the N-dimensional (k, l) block-matrix element associated with the k-th and l-th principal layers of the Green's function matrix

$$
\hat{\mathcal{G}}_T(E, \mathbf{k}_{\|}) = \left[E\hat{S}_T - \hat{H}_T \right]^{-1}.
\tag{B.18}
$$

We consider the scattering state in the case of the incident coming from the left electrode as illustrated in Fig. B.1. The scattering boundary conditions inside the electrodes,

$$
\psi(\mathbf{r}) =
\begin{cases}
\phi^{in}(\mathbf{r}) + \sum\limits_{i=1}^{N} r_i \phi_i^{ref}(\mathbf{r}) & \text{in the left electrode} \\
\sum\limits_{i=1}^{N} t_i \phi_i^{tra}(\mathbf{r}) & \text{in the right electrode},
\end{cases}
\tag{B.19}
$$

are transformed to

$$
C_l =
\begin{cases}
C_l^{in} + \sum\limits_{i=1}^{N} r_i C_{l,i}^{ref} & \text{for } l \leq 0 \\
\sum\limits_{i=1}^{N} t_i C_{l,i}^{tra} & \text{for } l \geq m+1,
\end{cases}
\tag{B.20}
$$

where C_l^{in} denotes the coefficient vector at the l-th principal layer associated with the incident propagating Bloch wave ϕ^{in}, and $C_{l,i}^A$ ($A = ref$ and tra) stand for the vectors related to the generalized Bloch states ϕ_i^A including evanescent waves. Here, the linear independence of the basis functions $\{\alpha_{l,\mu}\}$ at any principal layer l was assumed. As a consequence, proceeding along the same arguments as (6.13) – (6.27), we obtain the following wave-function matching formula from (B.17) and (B.20) [see Eq. (6.26)]:

$$
\begin{bmatrix}
\mathcal{W}_{0,0}R_0^{ref} - I & \mathcal{W}_{0,m+1}R_{m+2}^{tra} \\
\mathcal{W}_{m+1,0}R_0^{ref} & \mathcal{W}_{m+1,m+1}R_{m+2}^{tra} - I
\end{bmatrix}
\begin{bmatrix}
C_0 \\
C_{m+1}
\end{bmatrix}
$$

$$
= -\begin{bmatrix}
\mathcal{W}_{0,0} \\
\mathcal{W}_{m+1,0}
\end{bmatrix}
(C_{-1}^{in} - R_0^{ref}C_0^{in}) . \tag{B.21}
$$

Here,

$$
\begin{aligned}
\mathcal{W}_{0,0} &= \mathcal{G}_T(0,0)B_{-1}^{\dagger} \\
\mathcal{W}_{0,m+1} &= \mathcal{G}_T(0,m+1)B_{m+1} \\
\mathcal{W}_{m+1,0} &= \mathcal{G}_T(m+1,0)B_{-1}^{\dagger} \\
\mathcal{W}_{m+1,m+1} &= \mathcal{G}_T(m+1,m+1)B_{m+1}.
\end{aligned}
\tag{B.22}
$$

After all, the set of coefficient vectors $\{C_l\}$ corresponding to the scattering wave function $\psi(r)$ in the area overlying the transition region, $l = -1, 0, ..., m+1, m+2$, is calculated for an assigned set of incident coefficient vectors $\{C_{-1}^{in}, C_0^{in}\}$ by using (B.21) together with (B.16) and (B.17). It should be remarked that in determining the set $\{C_l\}$ by these equations, all of the block-matrix elements of $\hat{\mathcal{G}}_T(E, \boldsymbol{k}_{||})$ of (B.18) are not necessary, but only the left and right block columns, i.e., $\mathcal{G}_T(k, l)$ $(k = 0, 1, ..., m+1,$ and $l = 0, m+1)$, are needed. In this situation, when the transition region is that of a large-sized structure, it is appropriate to implement conventional iterative algorithms, such as steepest-descent and conjugate-gradient algorithms, for obtaining the columnar elements $\mathcal{G}_T(k, l)$ in question, since the relevant matrix $[E\hat{S}_T - \hat{H}_T]$ is Hermitian and these algorithms are expected to yield rapid convergence for simultaneous linear equations related to an Hermitian matrix.

B.3 Green's Function Formalism

Since the transition region is in contact with the electrodes via some connection area, the Green's function of the isolated transition region, $\hat{\mathcal{G}}_T(E, \boldsymbol{k}_{||})$ $(= [E\hat{S}_T - \hat{H}_T]^{-1})$ of (B.18), must be actually so modified into $\hat{G}_T^r(E, \boldsymbol{k}_{||})$, which is the retarded Green's function of the whole system portioned to the transition region, as to include the effects of the electrodes through the retarded self-energy terms $\hat{\Sigma}_{\{L,R\}}^r(E, \boldsymbol{k}_{||})$ as [cf. Eqs. (9.60) and (9.64)]

$$
\hat{G}_T^r(E, \boldsymbol{k}_{||}) = \left[E\hat{S}_T - \hat{H}_T - \hat{\Sigma}_L^r(E, \boldsymbol{k}_{||}) - \hat{\Sigma}_R^r(E, \boldsymbol{k}_{||})\right]^{-1} , \tag{B.23}
$$

and equivalently, in the form of Dyson's equation

$$\hat{G}_T^r(E, \mathbf{k}_{||}) = \hat{\mathcal{G}}_T(E, \mathbf{k}_{||})$$
$$+ \hat{\mathcal{G}}_T(E, \mathbf{k}_{||}) \left[\hat{\Sigma}_L^r(E, \mathbf{k}_{||}) + \hat{\Sigma}_R^r(E, \mathbf{k}_{||}) \right] \hat{G}_T^r(E, \mathbf{k}_{||}). \quad \text{(B.24)}$$

It is assumed that the TB Hamiltonian matrix of the whole system has the same structure as that indicated in Fig. 9.2; above all, the connection between the transition region and left (right) electrode is described by only one N-dimensional block matrix B_{-1} (B_{m+1}). This assumption leads to simplified forms for $\Sigma_{\{L,R\}}^r(E, \mathbf{k}_{||})$ as [see Eq. (9.67)]

$$\hat{\Sigma}_L^r(E, \mathbf{k}_{||}) = \begin{bmatrix} \Sigma_L^r(0) & 0 & \cdots & 0 \\ 0 & 0 & \cdots & 0 \\ \vdots & & & \vdots \\ 0 & 0 & \cdots & 0 \end{bmatrix}$$

$$\hat{\Sigma}_R^r(E, \mathbf{k}_{||}) = \begin{bmatrix} 0 & \cdots & 0 & 0 \\ \vdots & & & \vdots \\ 0 & \cdots & 0 & 0 \\ 0 & \cdots & 0 & \Sigma_R^r(m+1) \end{bmatrix}, \quad \text{(B.25)}$$

and moreover, explicit descriptions of the nonzero block matrix elements $\Sigma_L^r(0)$ and $\Sigma_R^r(m+1)$ as [see Eq. (9.70) in Theorem 9.2]

$$\Sigma_L^r(0) = B_{-1}^\dagger R_0^{ref} \quad \text{and} \quad \Sigma_R^r(m+1) = B_{m+1} R_{m+2}^{tra}. \quad \text{(B.26)}$$

Efficient and stable methods for calculating R^{ref} and R^{tra} have been presented in Section B.1. A feature of (B.26) is that it allows us to calculate the self-energy terms at a *purely* real energy without adding a small imaginary part to the energy.

Let $G_T^r(k, l)$ be the N-dimensional (k, l) block-matrix element associated with the k-th and l-th principal layers of $\hat{G}_T^r(E, \mathbf{k}_{||})$. From the consideration based on the Green's function matching in the OBM scheme [cf. Eq. (9.91)], one can find that the block-matrix elements G_T^r's at the four corners are expressed as [see Eqs. (9.93) – (9.96)]

$$G_T^r(0,0) = \tilde{\mathcal{G}}_T(0,0)\left[I - \sum_L^r(0)\tilde{\mathcal{G}}_T(0,0)\right]^{-1}$$

$$G_T^r(m+1,0) = \left[I - \mathcal{G}_T(m+1,m+1)\sum_R^r(m+1)\right]^{-1}$$
$$\times\, \mathcal{G}_T(m+1,0)\left[I - \sum_L^r(0)\tilde{\mathcal{G}}_T(0,0)\right]^{-1}$$

$$G_T^r(0,m+1) = \left[I - \mathcal{G}_T(0,0)\sum_L^r(0)\right]^{-1}\mathcal{G}_T(0,m+1)$$
$$\times\, \left[I - \sum_R^r(m+1)\tilde{\mathcal{G}}_T(m+1,m+1)\right]^{-1}$$

$$G_T^r(m+1,m+1) = \tilde{\mathcal{G}}_T(m+1,m+1)$$
$$\times\, \left[I - \sum_R^r(m+1)\tilde{\mathcal{G}}_T(m+1,m+1)\right]^{-1}, \text{(B.27)}$$

where

$$\tilde{\mathcal{G}}_T(0,0) = \mathcal{G}_T(0,0) + \mathcal{G}_T(0,m+1)\sum_R^r(m+1)$$
$$\times\, \left[I - \mathcal{G}_T(m+1,m+1)\sum_R^r(m+1)\right]^{-1}\mathcal{G}_T(m+1,0)$$

$$\tilde{\mathcal{G}}_T(m+1,m+1) = \mathcal{G}_T(m+1,m+1)$$
$$+\, \mathcal{G}_T(m+1,0)\sum_L^r(0)\left[I - \mathcal{G}_T(0,0)\sum_L^r(0)\right]^{-1}$$
$$\times\, \mathcal{G}_T(0,m+1). \quad \text{(B.28)}$$

After some algebra it is confirmed that G_T^r's represented by (B.27) and (B.28) are exact analytic solutions of Dyson's equation (B.24). Using $G_T^r(0,m+1)$ (or $G_T^r(m+1,0)$) thus determined, the electric conductance G is evaluated as, according to the NEGF formula [see Eq. (9.108) in Theorem 9.3],

$$G = \frac{2e^2}{h}\sum_{i,j}|t_{i,j}|^2\frac{v_i'}{v_j}$$
$$= \frac{2e^2}{h}\text{Tr}\left[\Gamma_L(0)G_T^a(0,m+1)\Gamma_R(m+1)G_T^r(m+1,0)\right]$$
$$= \frac{2e^2}{h}Tr\left[\Gamma_L(0)G_T^r(0,m+1)\Gamma_R(m+1)G_T^a(m+1,0)\right], \quad \text{(B.29)}$$

where

$$G_T^a(0,m+1) = G_T^r(m+1,0)^\dagger$$
$$G_T^a(m+1,0) = G_T^r(0,m+1)^\dagger \quad \text{(B.30)}$$

and

$$\Gamma_L(0) = i\left[\Sigma_L^r(0) - \Sigma_L^r(0)^\dagger\right]$$

$$\Gamma_R(m+1) = i\left[\Sigma_R^r(m+1) - \Sigma_R^r(m+1)^\dagger\right]. \tag{B.31}$$

As the size of the transition region becomes larger, directly inversing a matrix such as (B.23) turns out to be computationally harder. Obviously, it is very efficient to employ (B.27) and (B.28) in conductance calculations, since the dimension of the inverse matrices appearing in these equations is N, i.e., the number of atomic orbitals in *one* principal layer within the electrode regions, which should be compared with the $(m+2)N$ dimension of the matrices in (B.23), and furthermore, the relevant $\mathcal{G}_T(k,l)$ can be effectively calculated by iterative methods, thanks to the Hermiticity of $[E\hat{S}_T - \hat{H}_T]$ (see the last paragraph in Section B.2). Therefore, the potential power of the OBM scheme will be exhibited when it is applied to the exploration of physics for large-sized structures.

Finally, we summarize the relationships between the surface Green's functions and the coefficient vectors $\{C_l\}$ of the scattering wave at the surfaces of the electrodes [see Eqs. (9.104) and (9.105)]: For an incident wave from the left electrode whose coefficient vectors are $\{C_l^{in}\}$ ($l \leq 0$),

$$C_0 = iG_T^r(0,0)\Gamma_L(0)C_0^{in}$$

$$C_{m+1} = iG_T^r(m+1,0)\Gamma_L(0)C_0^{in}, \tag{B.32}$$

which is equivalent to the wave-function matching formula (B.21). For an incident wave from the right electrode whose coefficient vectors are $\{C_l^{in}\}$ ($l \geq m+1$),

$$C_0 = iG_T^r(0,m+1)\Gamma_R(m+1)C_{m+1}^{in}$$

$$C_{m+1} = iG_T^r(m+1,m+1)\Gamma_R(m+1)C_{m+1}^{in}. \tag{B.33}$$

When the number of atomic orbitals in any principal layer in the transition region is assumed to be N, equally to that in the electrode regions, (B.32) and (B.33) are extended as follows [see Eq. (9.152) in Theorem 9.5]: For an incident wave from the left electrode,

$$C_l = iG_T^r(l,0)\Gamma_L(0)C_0^{in} \quad (0 \leq l \leq m+1). \tag{B.34}$$

For an incident wave from the right electrode,

$$C_l = iG_T^r(l,m+1)\Gamma_R(m+1)C_{m+1}^{in} \quad (0 \leq l \leq m+1). \tag{B.35}$$

Bibliography

Abdurixit A., Baratoff A. and Galli G. (1999). Tight-binding linear scaling method: Applications to silicon surfaces. In *Modern Research in High-Technology and Science*, (eds.) Geni M. and Sheyhidin I. (Tokyo, Kansei Printing).

Agraït N., Levy Yeyati A. and van Ruitenbeek J.M. (2003). Quantum properties of atomic-sized conductors, *Phys. Rep.* **377**, 2–3, pp. 81–279.

Ami S., Hliwa M. and Joachim C. (2003). Molecular 'OR' and 'AND' logic gates integrated in a single molecule, *Chem. Phys. Lett.* **367**, pp. 662–668.

Aoki M. (1993). Rapidly convergent bond order expansion for atomistic simulations, *Phys. Rev. Lett.* **71**, 23, pp. 3842–3845.

Appelbaum J.A. and Blount E.I. (1973). Sum rule for crystalline metal surfaces, *Phys. Rev. B* **8**, 2, pp. 483–491.

Appelbaum J.A. and Hamann D.R. (1972). Self-consistent electronic structure of solid surfaces, *Phys. Rev. B* **6**, 6, pp. 2166–2177.

Arrhenius S. (1889). Ober die reacktionsgeschwindigkeit bei der inversion von rohrzucker durch säuren, *Z. Phys. Chem.* **4**, pp. 226–248.

Ayuela A., Raebiger H., Puska M.J. and Nieminen R.M. (2002). Spontaneous magnetization of aluminum nanowires deposited on the NaCl(100) surface, *Phys. Rev. B* **66**, 3, pp. 035417-1–035417-8.

Büttiker M., Imry Y., Landauer R. and Pinhas S. (1985). Generalized many-channel conductance formula with application to small rings, *Phys. Rev. B* **31**, 10, pp. 6207–6215.

Bachelet G.B., Hamann D.R. and Schlüter M. (1982). Pseudopotentials that work: From H to Pu, *Phys. Rev. B* **26**, 8, pp. 4199–4228.

Baer R. and Head-Gordon M. (1997). Chebyshev expansion methods for electronic structure calculations on large molecular systems, *J. Chem. Phys.* **107**, 23, pp. 10003–10013.

Barnett R.N. and Landman Uzi (1997). Cluster-derived structures and conductance fluctuations in nanowires, *nature* **387**, 19, pp. 788–791.

Barrio D.P., Glasser M.L., Velasco V.R. and García-Moliner F. (1989). Theory of quantum wells in external electric fields, *J. Phys.: Condens. Matter* **1**, 27, pp. 4339–4351.

Bates K.R., Daniels A.D. and Scuseria G.E. (1998). Comparison of conjugate gradient density matrix search and Chebyshev expansion methods for avoiding diagonalization in large-scale electronic structure calculations, *J. Chem. Phys.* **109**, 9, pp. 3308–3312.

Beck T.L. (1999). Multigrid High-order mesh refinement techniques for composite grid electrostatics calculations, *J. Comput. Chem.* **20**, 16, pp. 1731–1739.

Beck T.L. (2000). Real-space mesh techniques in density-functional theory, *Rev. Mod. Phys.* **72**, 4, pp. 1041–1080.

Becke A.D. (1986). Completely numerical calculations on diatomic molecules in the local-density approximation, *Phys. Rev. A* **33**, 4, pp. 2786–2788.

Becke A.D. (1988). A multicenter numerical integration scheme for polyatomic molecules, *J. Chem. Phys.* **88**, 4, pp. 2547–2553.

Becke A.D. (1992). Density-functional thermochemistry. II. The effect of the Perdew–Wang generalized-gradient correlation correction, *J. Chem. Phys.* **97**, 12, pp. 9173–9177.

Bennett A.J. and Duke C.B. (1967). Self-Consistent-Field Model of Bimetallic Interfaces.I.Dipole Effects, *Phys. Rev.* **160**, 3, pp. 541–553.

Bernholc J., Yi J.-Y. and Sullivan D.J. (1991). Structural transitions in metal clusters, *Faraday Discuss* **92**, pp. 217–228.

Bertsch G.F., Iwata J.-I., Rubio A. and Yabana K. (2000). Real-space, real-time method for the dielectric function, *Phys. Rev. B* **62**, 12, pp. 7998–8002.

Brandbyge M., Kobayashi N. and Tsukada M. (1999). Conduction channels at finite bias in single-atom gold contacts, *Phys. Rev. B* **60**, 24, pp. 17064–17070.

Brandbyge M., Mozos J.-L., Ordejón P., Taylor J. and Stokbro K. (2002). Density-functional method for nonequilibrium electron transport, *Phys. Rev. B* **65**, 16, pp. 165401-1–165401-17.

Brandt A. (1977). Multi-level adaptive solutions to boundary-value problems, *Math. Comput.* **31**, 138, pp. 333–390.

Brandt A., McCormick S.F. and Ruge J. (1983). Multigrid methods for differential eigenproblems, *J. Sci. Stat. Comput.* **4**, 2, pp. 244–260.

Briggs W.L. (1987). *A Multigrid Tutorial* (Philadelphia, SIAM).

Briggs E.L., Sullivan D.J. and Bernholc J. (1995). Large-scale electronic-structure calculations with multigrid acceleration, *Phys. Rev. B* **52**, 8, pp. R5471–R5474.

Briggs E.L., Sullivan D.J. and Bernholc J. (1996). Real-space multigrid-based approach to large-scale electronic structure calculations, *Phys. Rev. B* **54**, 20, pp. 14362–14375.

Buongiorno Nardelli M. (1999). Electronic transport in extended systems: Application to carbon nanotubes, *Phys. Rev. B* **60**, 11, pp. 7828–7833.

Buongiorno Nardelli M., Fattebert J.-L. and Bernholc J. (2001). $O(N)$ real-space method for *ab initio* quantum transport calculations: Application to carbon nanotube–metal contacts, *Phys. Rev. B* **64**, 24, pp. 245423-1–245423-5.

Bylander D.M., Kleinman L. and Lee S. (1990). Self-consistent calculations of the energy bands and bonding properties of $B_{12}C_3$, *Phys. Rev. B* **42**, 2, pp. 1394–1403.

Car R. and Parrinello M. (1985). Unified approach for molecular dynamics and density-functional theory, *Phys. Rev. Lett.* **55**, 22, pp. 2471–2474.

Carlsson A.E. (1995). Order-N density-matrix electronic-structure method for general potentials, *Phys. Rev. B* **51**, 20, pp. 13935–13941.

Caroli C., Combescot R., Nozieres P. and Saint-James D. (1971). Direct calculation of the tunneling current, *J. Phys. C* **4**, 8, pp. 916–929.

Cerdá J., Van Hove M.A., Sautet P. and Salmeron M. (1997). Efficient method for the simulation of STM images. I. Generalized Green-function formalism, *Phys. Rev. B* **56**, 24, pp. 15885–15899.

Challacombe M. (1999). A simplified density matrix minimization for linear scaling self-consistent field theory, *J. Chem. Phys.* **110**, 5, pp. 2332–2342.

Chelikowsky J.R., Jing X.,Wu K. and Saad Y. (1996). Molecular dynamics with quantum forces: Vibrational spectra of localized systems, *Phys. Rev. B* **53**, 18, pp. 12071–12079.

Chelikowsky J.R., Troullier N. and Saad Y. (1994a). Finite-difference-pseudopotential method: Electronic structure calculations without a basis, *Phys. Rev. Lett.* **72**, 8, pp. 1240–1243.

Chelikowsky J.R., Troullier N., Wu K. and Saad Y. (1994b). Higher-order finite-difference pseudopotential method: An application to diatomic molecules, *Phys. Rev. B* **50**, 16, pp. 11355–11364.

Chen Y.C. and Di Ventra M. (2003). Shot noise in nanoscale conductors from first principles, *Phys. Rev. B* **67**, 15, pp. 153304-1–153304-4.

Choi H.J. and Ihm J. (1999). Ab initio pseudopotential method for the calculation of conductance in quantum wires, *Phys. Rev. B* **59**, 3, pp. 2267–2275.

Cohen M.H. and Heine V. (1970). The fitting of pseudopotentials to experimental data and their subsequent application. *Solid State Physics, Vol. 24*, (Eds.) Ehrenreich H., Seitz F. and Turnbull D. (New York and London, Academic Press).

Costiner S. and Ta'asan S. (1995). Adaptive multigrid techniques for large-scale eigenvalue problems: Solutions of the Schrödinger problem in two and three dimensions, *Phys. Rev. E* **51**, 4, pp. 3704–3717.

Cuevas J.C., Levy Yeyati A., Martín-Rodero A., Rubio Bollinger G., Untiedt C. and Agraït N. (1998). Evolution of conducting channels in metallic atomic contacts under elastic deformation, *Phys. Rev. Lett.* **81**, 14, pp. 2990–2993.

Cuevas J.C., Martín-Rodero A. and Levy Yeyati A. (1996). Hamiltonian approach to the transport properties of superconducting quantum point contacts, *Phys. Rev. B* **54**, 10, pp. 7366–7379.

Damle P.S., Ghosh A.W. and Datta S. (2001). Unified description of molecular conduction: From molecules to metallic wires, *Phys. Rev. B* **64**, 20, pp. 201403-1–201403-4.

Datta S. (1995). *Electronic Transport in Mesoscopic Systems* (New York, Cambridge University Press).

Di Ventra M., Kim S.-G., Pantelides S.T. and Lang N.D. (2001). Temperature effects on the transport properties of molecules, *Phys. Rev. Lett.* **86**, 2, pp. 288–291.

Di Ventra M. and Lang N.D. (2001). Transport in nanoscale conductors from first principles, *Phys. Rev. B* **65**, 4, pp. 045402-1–045402-8.

Di Ventra M., Pantelides S.T. and Lang N.D. (2000). First-principles calculation of transport properties of a molecular device, *Phys. Rev. Lett.* **84**, 5, pp. 979–982.

Di Ventra M., Pantelides S.T. and Lang N.D. (2002). Current-induced forces in molecular wires, *Phys. Rev. Lett.* **88**, 4, pp. 046801-1–046801-4.

Doyen G., Drakova D. and Scheffler M. (1993). Green-function theory of scanning tunneling microscopy: Tunnel current and current density for clean metal surfaces, *Phys. Rev. B* **47**, 15, pp. 9778–9790.

Dreizler R.M. and da Providencia J. (1985). *Density Functional Methods in Physics*, (New York, Plenum).

Dunlap B.I. (1984). Explicit treatment of correlation within density-functional theories that use the kinetic-energy operator, *Phys. Rev. A* **29**, 5, pp. 2902–2905.

Dunlap B.I. (1988). Symmetry and spin density functional theory, *Chem. Phys.* **125**, 1, pp. 89–97.

Egami Y., Sasaki T., Tsukamoto S., Ono T. and Hirose K. (2004). First-principles study on electron conduction property of monatomic sodium nanowire, *Mater. Trans. JIM* **45**, 5, pp. 1433–1436.

Eigler D.M. and Schweizer E.K. (1990). Positioning single atoms with a scanning tunnelling microscope, *Nature* **344**, pp. 524–526.

Emberly E.G. and Kirczenow G. (1998). Theoretical study of electrical conduction through a molecule connected to metalic nanocontacts, *Phys. Rev. B* **58**, 16, pp. 10911–10920.

Emberly E.G. and Kirczenow G. (1999). Electron standing-wave formation in atomic wires, *Phys. Rev. B* **60**, 8, pp. 6028–6033.

Ewald P.P. (1921). Die berechnung optischer und elektrostatischer gitterpotentiale, *Ann. Phys.* **64**, pp. 253–287.

Fattebert J.-L. (1999). Finite difference schemes and block rayleigh quotient iteration for electronic structure calculations on composite grids, *J. Comput. Phys.* **149**, pp. 75–94.

Faulkner J.S., Wang Y. and Stocks G.M. (1995). Electrons in extended systems, *Phys. Rev. B* **52**, 24, pp. 17106–17111.

Fermi E. (1927). Un metodo statistice per la determinazione di alcune proprieta dell'atomo, *Rend. Accad. Lincei.* **6**, pp. 602–607.

Fermi E. (1928a). A statistical method for the determination of some atomic properties and the application of this method to the theory of the periodic system of elements, *Z. Phys.* **48**, pp. 73–79.

Fermi E. (1928b). Sulla deduzione statistica di alcune proprieta dell'atomo. Applicazione alla teoria del systema periodico degli elementi, *Rend. Accad. Lincei.* **7**, pp. 342–346.

Fernando G.W., Qian G.-X., Weinert M. and Davenport J.W. (1989). First-principles molecular dynamics for metals, *Phys. Rev. B* **40**, 11, pp. 7985–7988.

Ferrante J. and Smith J.R. (1985). Theory of the bimetallic interface, *Phys. Rev. B* **31**, 6, pp. 3427–3434.

Fisher D.S. and Lee P.A. (1981). Relation between conductivity and transmission matrix, *Phys. Rev. B* **23**, 12, pp. 6851–6854.

Fröhlich H. (1954). On the theory of superconductivity: The one dimensional case, *Proc. R. Soc. London, Ser. A* **223**, pp. 296–305.

Fujimoto Y. and Hirose K. (2003a). First-principles calculation method of electron-transport properties of metallic nanowires, *Nanotechnology* **14**, 2, pp. 147–151.

Fujimoto Y. and Hirose K. (2003b). First-principles treatments of electron transport properties for nanoscale junctions, *Phys. Rev. B* **67**, 19, pp. 195315-1–195315-12.

Fujimoto Y. and Hirose K. (2004). Erratum: First-principles treatments of electron transport properties for nanoscale junctions, *Phys. Rev. B* **69**, 11, pp. 119901-1–119901-1.

Galli G. (2000). Large-scale electronic structure calculations using linear scaling methods, *Phys. Stat. Sol. B* **217**, pp. 231–249.

Galli G. and Mauri F. (1994). Large scale quantum simulations: C_{60} impacts on a semiconducting surface, *Phys. Rev. Lett.* **73**, 25, pp. 3471–3474.

Galli G. and Parrinello M. (1992). Large scale electronic structure calculations, *Phys. Rev. Lett.* **69**, 24, pp. 3547–3550.

García-Moliner F. and Velasco V.R. (1991). Matching methods for single and multiple interfaces: Discrete and continuous media, *Phys. Rep.* **200**, 3, pp. 83–125.

García-Moliner F. and Velasco V.R. (1992). *Theory of Single and Multiple Interfaces*, Singapore, World Scientific.

García-Moliner F., Pérez-Álvarez R., Rodriguez-Coppola H. and Velasco V.R. (1990). A general theory of matching for layered systems, *J. Phys. A: Math. Gen.* **23**, 8, pp. 1405–1420.

Gianturco F.A., Kashenock G.Y., Lucchese R.R. and Sanna N. (2002). Low-energy resonant structures in electron scattering from C_{20} fullerene, *J. Chem. Phys.* **116**, 7, pp. 2811–2824.

Goedecker S., Teter M. and Hutter J. (1996). Separable dual-space Gaussian pseudopotentials, *Phys. Rev. B* **54**, 3, pp. 1703–1710.

Gohda Y., Nakamura Y., Watanabe K. and Watanabe S. (2000). Self-consistent density functional calculation of field emission currents from metals, *Phys. Rev. Lett.* **85**, 8, pp. 1750–1753.

Guillermet A.F. and Grimvall G. (1989). Homology of interatomic forces and Debye temperatures in transition metals, *Phys. Rev. B* **40**, 3, pp. 1521–1527.

Guinea F., Tejedor C., Flores F. and Louis E. (1983). Effective two-dimensional Hamiltonian at surfaces, *Phys. Rev. B* **28**, 8, pp. 4397–4402.

Gutknecht M.H. (1993). Variants of Bi-CGSTAB for matrices with complex spectrum, *SIAM J. Sci. Comput.* **14**, pp. 1020–1033.

Gygi F. and Galli G. (1995). Real-space adaptive-coordinate electronic-structure calculations, *Phys. Rev. B* **52**, 4, pp. R2229–R2232.

Hackbusch W. (1985). *Multi-Grid Methods and Applications* (New York, Springer-Verlag).

Hamada N., Sawada S. and Oshiyama A. (1992). New one-dimensional conductors: Graphitic microtubules, *Phys. Rev. Lett.* **68**, 10, pp. 1579–1581.

Hamann D.R. (1989). Generalized norm-conserving pseudopotentials, *Phys. Rev. B* **40**, 5, pp. 2980–2987.

Hamann D.R., Schlüter M. and Chiang C. (1979). Norm-conserving pseudopotentials, *Phys. Rev. Lett.* **43**, 20, pp. 1494–1497.

Harris J. (1984), Adiabatic-connection approach to Kohn–Sham theory, *Phys. Rev. A* **29**, 4, pp. 1648–1659.

Havu P., Torsti T., Puska M.J. and Nieminen R.M. (2002). Conductance oscillations in metallic nanocontacts, *Phys. Rev. B* **66**, 7, pp. 075401-1–075401-5.

Hernández E. and Gillan M.J. (1995). Self-consistent first-principles technique with linear scaling, *Phys. Rev. B* **51**, 15, pp. 10157–10160.

Hernández E., Gillan M.J. and Goringe C.M. (1996). Linear-scaling density-functional-theory technique: The density-matrix approach, *Phys. Rev. B* **53**, 11, pp. 7147–7157.

Hestenes M.R. and Stiefel E. (1952). Methods of conjugate gradients for solving linear systems, *J. Res. Nat. Bur. Stand.* **49**, 6, pp. 409–436.

Hierse W. and Stechel E.B. (1994). Order-N methods in self-consistent density-functional calculations, *Phys. Rev. B* **50**, 24, pp. 17811–17819.

Hirose K., Kobayashi N. and Tsukada M. (2004). *Ab initio* calculations for quantum transport through atomic bridges by the recuirsion transfer-matrix method, *Phys. Rev. B* **69**, 24, pp. 245412-1–245412-5.

Hirose K. and Ono T. (2001). Direct minimization to generate electronic states with proper occupation numbers, *Phys. Rev. B* **64**, 8, pp. 085105-1–085105-5.

Hirose K. and Tsukada M. (1994) First-principles theory of atom extraction by scanning tunneling microscopy, *Phys. Rev. Lett.* **73**, 1, pp. 150–153.

Hirose K. and Tsukada M. (1995). First-principles calculation of the electoronic structure for a bielectrode junction system under strong field and current, *Phys. Rev. B* **51**, 8, pp. 5278–5290.

Hohenberg P. and Kohn W. (1964). Inhomogeneous electron gas, *Phys. Rev.* **136**, 3B, pp. B864–B871.

Holzwarth N.A.W. and Lee M.J.G. (1978). Surface electronic wave functions of a semi-infinite muffin-tin lattice. I. The spherical-wave method, *Phys. Rev. B* **18**, 10, pp. 5350–5364.

Holzwarth N.A.W., Matthews G.E., Dunning R.B., Tackett A.R. and Zeng,Y. (1997). Comparison of the projector augmented-wave, pseudopotential and linearized augmented-plane-wave formalisms for density-functional calculations of solids, *Phys. Rev. B* **55**, 4, pp. 2005–2017.

Hori T., Takahashi H. and Nitta T. (2003a). Hybrid QM/MM molecular dynamics simulations for an ionic S_N2 reaction in the supercritical water: $OH^- + CH_3Cl \rightarrow CH_3OH + Cl^-$, *J. Comput. Chem.* **24**, 2, pp. 209–221.

Hori T., Takahashi H. and Nitta T. (2003b). Hybrid quantum chemical studies for the methanol formation reaction assisted by the proton transfer mechanism in supercritical water: $CH_3Cl + nH_2O \rightarrow CH_3OH + HCl + (n-1)H_2O$, *J. Chem. Phys.* **119**, 16, pp. 8492–8499.

Hoshi T., Arai M. and Fujiwara T. (1995). Density-functional molecular dynamics with real-space finite difference, *Phys. Rev. B* **52**, 8, pp. R5459–R5462.

Hoshi T. and Fujiwara T. (2000). Theory of composite-band Wannier states and order-N electronic-structure calculations, *J. Phys. Soc. Jpn.* **69**, 12, pp. 3773–3776.

Hoshi T. and Fujiwara T. (2003). Dynamical Brittle Fractures of Nanocrystalline Silicon using Large-Scale Electronic Structure Calculations, *J. Phys. Soc. Jpn.* **72**, 10, pp. 2429–2432.

Huber K.P. and Herzberg G. (1979). *Constants of Diatomic Molecules*, New York, Van Nostrand.

Hummel W. and Bross H. (1998). Determining the electronic properties of semi-infinite crystals, *Phys. Rev. B* **58**, 3, pp. 1620–1632.

Ichimura M., Kusakabe K., Watanabe S. and Onogi T. (1998). Flat-band ferromagnetism in extended Delta-chain Hubbard models, *Phys. Rev. B* **58**, 15, pp. 9595–9598.

Iijima S. (1991). Helical microtubules of graphitic carbon, *Nature* **354**, pp. 56–58.

Imry Y. (1997). *Introduction to Mesoscopic Physics*, New York, Oxford University Press.

Itoh S., Ordejón P., Drabold D.A. and Martin R.M. (1996). Structure and energetics of giant fullerenes: An order-N molecular-dynamics study, *Phys. Rev. B* **53**, 4, pp. 2132–2140.

Itoh S., Ordejón P. and Martin R.M. (1995). Order-N tight-binding molecular dynamics on parallel computers, *Compt. Phys. Comn.* **88**, pp. 173–185.

Ivanov V.K., Kashenock G.Yu., Polozkov R.G. and Solov'yov A.V. (2001). Photoionization cross sections of the fullerenes C_{20} and C_{60} calculated in a simple spherical model, *J. Phys. B* **34**, 21, pp. L669–L677.

Janak J.F. (1978). Proof that $\partial E/\partial n_i = \varepsilon$ in density-functional theory, *Phys. Rev. B* **18**, 12, pp. 7165–7168.

Jauho A.-P., Wingreen N.S. and Meir Y. (1994). Time-dependent transport in interacting and noninteracting resonant-tunneling systems, *Phys. Rev. B* **50**, 8, pp. 5528–5544.

Jensen F. (1999). *Introduction to Computational Chemistry*, (New York, Wiley).

Jin Y.-G., Jeong J.-W. and Chang K.J. (1999). Real-space electronic structure calculations of charged clusters and defects in semiconductors using a multi grid method., *Physica B* **273-274**, pp. 1003–1006.

Jing X., Troullier N., Dean D., Binggeli N., Chelikowsky J.R., Wu K. and Saad Y. (1994). *Ab initio* molecular-dynamics simulations of Si clusters using the higher-order finite-difference-pseudopotential method, *Phys. Rev. B* **50**, 16, pp. 12234–12237.

Joachim C., Gimzewski J.K., Schlittler R.R. and Chavy C. (1995). Electronic transparence of a single C_{60} molecule, *Phys. Rev. Lett.* **74**, 11, pp. 2102–2105.

Joachim C. and Gimzewski J.K. (1997). An electromechanical amplifier using a single molecule, *Chem. Phys. Lett.* **265**, pp. 353–357.

Joannopoulos J.D., Starkloff Th. and Kastner M. (1977). Theory of pressure dependence of the density of states and reflectivity of selenium, *Phys. Rev.*

Lett. **38**, 12, pp. 660–663.

Jones R.O. and Gunnarsson O. (1989). The density functional formalism, its applications and prospects, *Rev. Mod. Phys.* **61**, 3, pp. 689–746.

Kawabata A. (2000). Theory of electric conduction: The Landauer formula, *Butsuri* (in Japanese) **55**, 4, pp. 256–263.

Kawai T. and Watanabe K. (1996). Diffusion of a Si adatom on the Si(100) surface in an electric field, *Surf. Sci.* **357-358**, pp. 830–834.

Kawai T. and Watanabe K. (1997). Vibration and diffusion of surface atoms in strong electric fields, *Surf. Sci.* **382**, pp. 320–325.

Kawai T. Watanabe K. and Kobayashi K. (1998). Ab initio study on interaction between carbon atom and Si(100) surface in strong electric fields, *Ultramicroscopy* **73**, pp. 205–210.

Keldysh L.V. (1964). Diagram technique for nonequilibrium processes, *Zh. Eksp. Teor. Fiz.* **47**, pp. 1515–1527. Engl. trans.: *Sov. Phys. JETP* **20**, pp. 1018–1026(1965).

Kerker G.P. (1980). Non-singular atomic pseudopotentials for solid state applications, *J. Phys. C* **13**, 9, pp. L189–L194.

Kim J., Mauri F. and Galli G. (1995). Total-energy global optimizations using nonorthogonal localized orbitals, *Phys. Rev. B* **52**, 3, pp. 1640–1648.

Kittel C. (1986). *Introduction to solid state physics*, 6th ed. (New York, Wiley).

Kizuka T., Umehara S. and Fujisawa S. (2001). Metal-insulator transition in stable one-dimensional arrangements of single gold atoms, *Jpn. J. Appl. Phys.* **40**, 1A/B, pp. L71–L74.

Kleinman L. and Bylander D.M. (1982). Efficacious form for model pseudopotentials, *Phys. Rev. Lett.* **48**, 20, pp. 1425–1428.

Kobayashi K. (1999a). Norm-conserving pseudopotential database (NCPS97), *Comput. Mater. Sci.* **14**, 1-4, pp. 72–76.

Kobayashi K. (1999b). Tunneling into Bloch states from a tip in scanning tunneling microscopy, *Phys. Rev. B* **59**, 20, pp. 13251–13257.

Kobayashi N., Aono M. and Tsukada M. (2001). Conduction channels of Al wires at finite bias, *Phys. Rev. B* **64**, 12, pp. 121402-1–121402-4.

Kobayashi N., Brandbyge M. and Tsukada M. (2000). First-principles study of electron transport through monatomic Al and Na wires, *Phys. Rev. B* **62**, 12, pp. 8430–8437.

Kobayashi N. and Tsukada M. (1999). Numerical method for local density of states and current density decomposed into eigenchannels in multichannel system, *Jpn. J. Appl. Phys.* **38**, 6B, pp. 3805–3808.

Kohn W. and Sham L.J. (1965). Self-consistent equations including exchange and correlation effects, *Phys. Rev.* **140**, 4A, pp. A1133–A1138.

Kondo Y. and Takayanagi K. (2000). Synthesis and characterization of helical multi-shell gold nanowires, *Science* **289**, pp. 606–608.

Krans J.M., Muller C.J., Yanson I.K., Govaert Th.C.M., Hesper R. and van Ruitenbeek J.M. (1993). One-atom point contacts, *Phys. Rev. B* **48**, 19, pp. 14721–14724.

Krans J.M., Ruitenbeek J.M. and de Jongh L.J. (1996). Atomic structure and quantized conductance in metal point contacts, *Phys. B* **218**, pp. 228–233.

Krans J.M., Ruitenbeek J.M., Fisun V.V., Yanson I.K. and de Jongh L.J. (1995). The signature of conductance quantization in metallic point contacts, *Nature* **375**, pp. 767–769.

Kryachko E.S. and Ludena E.V. (1990). *Energy Density Functional Theory of Many Electron Systems*, (Boston, Kluwer Academic).

Kubo R. (1957). Statistical-mechanical theory of irreversible processes. I. General theory and simple applications to magnetic and conduction problems, *J. Phys. Soc. Jpn.* **12**, 6, pp. 570–586.

Lópes Sancho M.P., Lópes Sancho J.M. and Rubio J. (1984). Quick iterative scheme for the calculation of transfer matrices: Application to Mo (100), *J. Phys. F: Met. Phys.* **14**, 5, pp. 1205–1215.

Lópes Sancho M.P., Lópes Sancho J.M. and Rubio J. (1985). Highly convergent schemes for the calculation of bulk and surface Green functions, *J. Phys. F: Met. Phys.* **15**, 4, pp. 851–858.

Lam P.K. and Cohen M.L. (1981). *Ab initio* calculation of the static structural properties of Al, *Phys. Rev. B* **24**, 8, pp. 4224–4229.

Landauer R. (1957). Spatial variation of currents and fields due to localized scatterers in metallic conduction, *IBM J. Res. Dev.* **1**, pp. 223–231.

Landauer R. (1970). Electrical resistance of disordered one-dimensional lattices, *Phil. Mag.* **21**, pp. 863–867.

Lang N.D. (1992). Field-induced transfer of an atom between two closely spaced electrodes, *Phys. Rev. B* **45**, 23, pp. 13599–13606.

Lang N.D. (1995). Resistance of atomic wires, *Phys. Rev. B* **52**, 7, pp. 5335–5342.

Lang N.D. (1997a). Anomalous dependence of resistance on length in atomic wires, *Phys. Rev. Lett.* **79**, 7, pp. 1357–1360.

Lang N.D. (1997b). Conduction through single Mg and Na atoms linking two macroscopic electrodes, *Phys. Rev. B* **55**, 7, pp. 4113–4116.

Lang N.D. (1997c). Negative differential resistance at atomic contacts, *Phys. Rev. B* **55**, 15, pp. 9364–9366.

Lang N.D. and Avouris Ph. (1998). Oscillatory conductance of carbon-atom wires, *Phys. Rev. Lett.* **81**, 16, pp. 3515–3518.

Lang N.D. and Avouris Ph. (2000). Carbon-atom wires: Charge-transfer doping, voltage drop, and the effect of distortions, *Phys. Rev. Lett.* **84**, 2, pp. 358–361.

Lang N.D. and Avouris Ph. (2001). Electrical conductance of individual molecules, *Phys. Rev. B* **64**, 12, pp. 125323-1–125323-7.

Lang N.D. and Kohn W. (1970). Theory of metal surfaces:Charge density and surface energy, *Phys. Rev. B* **1**, 12, pp. 4555–4568.

Lang N.D. and Di Ventra M. (2003). Comment on "First-principles treatments of electron transport properties for nanoscale junctions", *Phys. Rev. B* **68**, 15, pp. 157301-1–157301-1.

Lang N.D. and Williams A.R. (1975). Self-consistent theory of the chemisorption of H, Li, and O on a metal surface, *Phys. Rev. Lett.* **34**, 9, pp. 531–534.

Lang N.D. and Williams A.R. (1978). Theory of atomic chemisorption on simple metals, *Phys. Rev. B* **18**, 2, pp. 616–636.

Larade B., Taylor J., Mehrez H. and Guo H. (2001). Conductance, *I-V* curves, and negative differential resistance of carbon atomic wires, *Phys. Rev. B* **64**, 7, pp. 075420-1–075420-10.

Larade B. and Bratkovsky A.M. (2003). Current rectification by simple molecular quantum dots: An *ab initio* study, *Phys. Rev. B* **68**, 23, pp. 235305-1–235305-8.

Lee D.H. and Joannopoulos J.D. (1981a). Simple scheme for surface-band calculations. I, *Phys. Rev. B* **23**, 10, pp. 4988–4996.

Lee D.H. and Joannopoulos J.D. (1981b). Simple scheme for surface-band calculations. II. The Green's function, *Phys. Rev. B* **23**, 10, pp. 4997–5004.

Lee H.-W., Sim H.-S., Kim D.-H. and Chang K.J. (2003). Towards unified understanding of conductance of stretched monatomic contacts, *Phys. Rev. B* **68**, 7, pp. 075424-1–075424-5.

Lent C.S. and Kirkner D.J. (1990). The quantum transmitting boundary method, *J. Appl. Phys.* **67**, 10, pp. 6353–6359.

Levy Yeyati A. (1992). Nonlinear conductance fluctuations in quantum wires: Appearance of two different energy scales, *Phys. Rev. B* **45**, 24, pp. 14189–14196.

Li X.-P., Nunes R.W. and Vanderbilt D. (1993). Density-matrix electronic-structure method with linear system-size scaling, *Phys. Rev. B* **47**, 16, pp. 10891–10894.

Liang G.C., Ghosh A.W., Paulsson M. and Datta S. (2004). Electrostatic potential profiles of molecular conductors, *Phys. Rev. B* **69**, 11, pp. 115302-1–115302-12.

Lipavský P., Spicka V. and Velický B. (1986). Generalized Kadanoff–Baym ansatz for deriving quantum transport equations, *Phys. Rev. B* **34**, 10, pp. 6933–6942.

Liu Y. and Guo H. (2004). Current distribution in B- and N-doped carbon nanotubes, *Phys. Rev. B* **69**, 11, pp. 115401-1–115401-6.

Long W., Sun Q-F., Guo H. and Wang J. (2003). Gate-controllable spin battery, *Appl. Phys. Lett.* **83**, 7, pp. 1397–1399.

Lyo I.W. and Avouris Ph. (1989). Negative differential resistance on the atomic scale: Implications for atomic scale devices, *Science* **245**, pp. 1369–1371.

Lyo I.-W. and Avouris P. (1991). Field-induced nanometer- to atomic-scale manipulation of silicon surfaces with the STM, Science **253**, pp. 173–176.

March N.H. (1975). *Self-Consistent Fields in Atoms*, (New York, Pergamon).

Marcus P.M. and Jepsen D.W. (1968). Accurate calculation of low-energy electron-diffraction intensities by the propagation-matrix method, *Phys. Rev. Lett.* **20**, 17, pp. 925–929.

Maria L.D. and Springborg M. (2000). Electronic structure and dimerization of a single monatomic gold wire, *Chem. Phys. Lett.* **323**, pp. 293-299.

Martín-Moreno L. and Vergés J.A. (1990). Random-Bethe-lattice model applied to the electronic structure of amorphous and liquid silicon, *Phys. Rev. B* **42**, 11, pp. 7193–7203.

Mauri F. and Galli G. (1994). Electronic-structure calculations and molecular-dynamics simulations with linear system-size scaling, *Phys. Rev. B* **50**, 7, pp. 4316–4326.

Mauri F., Galli G. and Car R. (1993). Orbital formulation for electronic-structure calculations with linear system-size scaling, *Phys. Rev. B* **47**, 15, pp. 9973–9976.

McLennan M.J., Lee Y. and Datta S. (1991). Voltage drop in mesoscopic systems: A numerical study using a quantum kinetic equation, *Phys. Rev. B* **43**, 17, pp. 13846–13884.

Mehl M.J. (2000). Occupation-number broadening schemes: Choice of "temperature", *Phys. Rev. B* **61**, 3, pp. 1654–1657.

Meir Y. and Wingreen N.S. (1992). Landauer formula for the current through an interacting electron region, *Phys. Rev. Lett.* **68**, 16, pp. 2512–2515.

Millam J.M. and Scuseria G.E. (1997). Linear scaling conjugate gradient density matrix search as an alternative to diagonalization for first principles electronic structure calculations, *J. Chem. Phys.* **106**, 13, pp. 5569–5577.

Mintmire J.W., Dunlap B.I. and White C.T. (1992). Are fullerene tubles metallic?, *Phys. Rev. Lett.* **68**, 5, pp. 631–634.

Mintmire J.W. and White C.T. (1998). First-principles band structures of armchair nanotubes, *Appl. Phys. A* **67**, pp. 65–69.

Miyamoto Y. and Saito M. (2001). Condensed phases of all-pentagon C_{20} cages as possible superconductors, *Phys. Rev. B* **63**, 16, pp. 161401-1–161401-4.

Mizobata J., Fujii A., Kurokawa S. and Sakai A. (2003a). Conductance of Al nanocontacts under high biases, *Jpn. J. Appl. Phys.* **42**, 7B, pp. 4680–4683.

Mizobata J., Fujii A., Kurokawa S. and Sakai A. (2003b). High-bias conductance of atom-sized Al contacts, *Phys. Rev. B* **68**, 15, pp. 155428-1–155428-7.

Modine N.A., Zumbach G. and Kaxiras E. (1997). Adaptive-coordinate real-space electronic-structure calculations for atoms, molecules, and solids, *Phys. Rev. B* **55**, 16, pp. 10289–10301.

Moroni E.G., Kresse G., Hafner J. and Furthmüller J. (1997). Ultrasoft pseudopotentials applied to magnetic Fe, Co, and Ni: From atoms to solids, *Phys. Rev. B* **56**, 24, pp. 15629–15646.

Mozos J.-L., Wan C.C., Taraschi G., Wang J. and Guo H. (1997). Quantized conductance of Si atomic wires, *Phys, Rev. B* **56**, 8, pp. R4351–R4354.

Nakamura A., Brandbyge M., Hansen L.B. and Jacobsen K.W. (1999). Density Functional Simulation of a Breaking Nanowire, *Phys. Rev. Lett.* **82**, 7, pp. 1538–1541.

Nakanishi S. and Tsukada M. (2001). Quantum loop current in a C_{60} molecular bridge, *Phys. Rev. Lett.* **87**, 12, pp. 126801-1–126801-4.

Needels M., Rappe A.M., Bristowe P.D. and Joannopoulos J.D. (1992). *Ab initio* study of a grain boundary in gold, *Phys. Rev. B* **46**, 15, pp. 9768–9771.

Nomura S., Iitaka T., Zhao X., Sugano T. and Aoyagi Y. (1998). Electronic structure of nanocrystalline/amorphous silicon: A novel quantum size effect, *Mater. Sci. Eng. B* **51**, pp. 146–149.

Nomura S., Iitaka T., Zhao X., Sugano T. and Aoyagi Y. (1999). Quantum-size effect in model nanocrystalline/amorphous mixed-phase silicon structures, *Phys. Rev. B* **59**, 15, pp. 10309–10314.

Nomura S., Zhao X., Aoyagi Y. and Sugano T. (1996). Electronic structure of a model nanocrystalline/amorphous mixed-phase silicon, *Phys. Rev. B* **54**,

19, pp. 13974–13979.

Northrup J. E., Yin M. T. and Cohen M. L. (1983). Pseudopotential local-spin-density calculations for Si_2, *Phys. Rev. A* **28**, 4, pp. 1945–1950.

Nunes R.W. and Vanderbilt D. (1994). Generalization of the density-matrix method to a nonorthogonal basis, *Phys. Rev. B* **50**, 23, pp. 17611–17614.

Ohnishi H., Kondo Y. and Takayanagi K. (1998). Quantized conductance through individual rows of suspended gold atoms, *Nature* **395**, pp. 780–783.

Okada S. and Oshiyama A. (2000). Magnetic ordering of Ga wires on Si(100) surfaces, *Phys. Rev. B* **62**, 20, pp. R13286–R13289.

Okamoto M. and Takayanagi K. (1999). Structure and conductance of a gold atomic chain, *Phys. Rev. B* **60**, 11, pp. 7808–7811.

Okano S., Shiraishi K. and Oshiyama A. (2004). Density-functional calculations and eigenchannel analyses for electron transport in Al and Si atomic wires, *Phys. Rev. B* **69**, 4, pp. 045401-1–045401-10.

Ono T. and Hirose K. (1999). Timesaving double-grid method for real-space electronic-structure calculations, *Phys. Rev. Lett.* **82**, 25, pp. 5016–5019.

Ono T. and Hirose K. (2003). First-principles study of Peierls instability in infinite single row Al wires, *Phys. Rev. B* **68**, 4, pp. 045409-1–045409-5.

Ono T. and Hirose K. (2004a). First-principles study on field evaporation for silicon atom on Si(001) surface, *J. Appl. Phys.* **95**, 3, pp. 1568–1571.

Ono T. and Hirose K. (2004b). Geometry and conductance of Al wires suspended between semi-infinite crystalline electrodes, *Phys. Rev. B* **70**, 3, pp. 033403-1–033403-4.

Ono T., Tsukamoto S. and Hirose K. (2003a). Magnetic orderings in Al nanowires suspended between electrodes, *Appl. Phys. Lett.* **82**, 25, pp. 4570–4572.

Ono T., Yamasaki H., Egami Y. and Hirose K. (2003b). A coherent relation between structure and conduction of infinite atomic wires, *Nanotechnology* **14**, pp. 299–303.

Ordejón P., Artacho E. and Soler J.M. (1996). Self-consistent order-N density-functional calculations for very large systems, *Phys. Rev. B* **53**, 16, pp. R10441–R10444.

Ordejón P., Drabold D.A., Grumbach M.P. and Martin R.M. (1993). Unconstrained minimization approach for electronic computations that scales linearly with system size, *Phys. Rev. B* **48**, 19, pp. 14646–14649.

Ordejón P., Drabold D.A., Martin R.M. and Grumbach M.P. (1995a). Linear system-size scaling methods for electronic-structure calculations, *Phys. Rev. B* **51**, 3, pp. 1456–1476.

Ordejón P., Drabold D.A., Martin R.M. and Itoh S. (1995b). Linear scaling method for phonon calculations from electronic structure, *Phys. Rev. Lett.* **75**, 7, pp. 1324–1327.

Oshima Y., Mouri K., Hirayama H. and Takayanagi K. (2004). Electronic conductance of gold helical multi-shell nanowire, *Phys. Rev. Lett.*, to be published.

Otani M., Ono T. and Hirose K. (2004). First-principles study of electron transport through C_{20} cages, *Phys. Rev. B* **69**, 12, pp. 121408-1–121408-4.

Ozaki T. (2001). Efficient recursion method for inverting an overlap matrix, *Phys. Rev. B* **64**, 19, pp. 195110-1–195110-7.

Ozaki T. Aoki M. and Pettifor D.G. (2000). Block bond-order potential as a convergent moments-based method, *Phys. Rev. B* **61**, 12, pp. 7972–7988.

Ozaki T. and Terakura K. (2001). Convergent recursive $O(N)$ method for *ab initio* tight-binding calculations, *Phys. Rev. B* **64**, 19, pp. 195126-1–195126-4.

Palacios J.J., Pérez-Jiménez A.J., Louis E., SanFabián E. and Vergés J.A. (2002). First-principles approach to electrical transport in atomic-scale nanostructures, *Phys. Rev. B* **66**, 3, pp. 035322-1–035322-14.

Parr R.G. and Yang W. (1989). *Density-Functional Theory of Atoms and Molecules*, (New York, Oxford University Press).

Pascual J.I., Méndez J., Gómez-Herrero J., Baró A.M., García N. and Binh V.T. (1993). Quantum contact in gold nanostructures by scanning tunneling microscopy, *Phys. Rev. Lett.* **71**, 12, pp. 1852–1855.

Payne M.C., Teter M.P., Allan D.C., Arias T.A. and Joannopoulos J.D. (1992). Iterative minimization techniques for *ab initio* total-energy calculations: Molecular dynamics and conjugate gradients, *Rev. Mod. Phys.* **64**, 4, pp. 1045–1097.

Pederson M. R. and Jackson K. A. (1991). Pseudoenergies for simulations on metallic systems, *Phys. Rev. B* **43**, 9, pp. 7312–7315.

Peierls R.E. (1955). *Quantum Theory of Solid* (London, Oxford University Press).

Perdew J.P., Chevary J.A., Vosko S.H., Jackson K.A., Pederson M.R., Singh D.J. and Fiolhais C. (1992). Atoms, molecules, solids, and surfaces: Applications of the generalized gradient approximation for exchange and correlation, *Phys. Rev. B* **46**, 11, pp. 6671–6687.

Perdew J.P. and Zunger A. (1981). Self-interaction correction to density-functional approximations for many-electron systems, *Phys. Rev. B* **23**, 10, pp. 5048–5079.

Pernas P.L., Martín-Rodero A. and Flores F. (1990). Electrochemical-potential variations across a constriction, *Phys. Rev. B* **41**, 12, pp. 8553–8556.

Phillips J.C. (1958). Energy-band interpolation scheme based on a pseudopotential, *Phys. Rev.* **112**, 3, pp. 685–695.

Prinzbach H., Weilwer A., Landenberger P., Wahl F., Wörth J., Scott L.T., Gelmont M., Olevano D. and Issendorff B.v. (2000). Gas-phase production and photoelectron spectroscopy of the smallest fullerene, C_{20}, *Nature* **407**, pp. 60–63.

Redondo A., Goddard W.A. and McGill T.C. (1977). *Ab initio* effective potentials for silicon, *Phys. Rev. B* **15**, 10, pp. 5038–5048.

Roche S. (1999). Quantum transport by means of $O(N)$ real-space methods, *Phys. Rev. B* **59**, 3, pp. 2284–2291.

Roland C., Larade B., Taylor J. and Guo H. (2001). *Ab initio* I–V characteristics of short C_{20} chains, *Phys. Rev. B* **65**, 4, pp. 041401-1–041401-4.

Rubio G., Agraït N. and Vieira S. (1996). Atomic-sized metallic contacts: Mechanical properties and electronic transport, *Phys. Rev. Lett.* **76**, 13, pp. 2302–2305.

Sánchez-Portal D., Artacho E., Junquera J., Ordejón P., García A. and Soler J.M. (1999). Stiff monatomic gold wires with a spinning zigzag geometry, *Phys. Rev. Lett.* **83**, 19, pp. 3884–3887.

Saito R., Fujita M., Dresselhaus G. and Dresselhaus M.S. (1992). Electronic structure of chiral graphene tubules, *Appl. Phys. Lett.* **60**, 18, pp. 2204–2206.

Sanvito S., Lambert C.J., Jefferson J.H. and Bratkovsky A.M. (1999). General Green's-function formalism for transport calculations with *spd* Hamiltonians and giant magnetoresistance in Co- and Ni-based magnetic multilayers, *Phys. Rev. B* **59**, 18, pp. 11936–11948.

Sasaki T., Egami Y., Tanide A., Ono T., Goto H. and Hirose K. (2004). First-principles calculation method for electronic structures of nanojunctions suspended between semi-infinite electrodes, *Mater. Trans. JIM* **45**, 5, pp. 1419–1421.

Sautet P. and Joachim C. (1988). Electronic transmission coefficient for the single-impurity problem in the scattering-matrix approach, *Phys. Rev. B* **38**, 17, pp. 12238–12247.

Scheer E., Joyez P., Esteve D., Urbina C. and Devoret M.H. (1997). Conduction channel transmissions of atomic-size aluminum contacts, *Phys. Rev. Lett.* **78**, 18, pp. 3535–3538.

Seitsonen A.P., Puska M.J. and Nieminen R.M. (1995). Real-space electronic-structure calculations: Combination of the finite-difference and conjugate-gradient methods, *Phys. Rev. B* **51**, 20, pp. 14057–14061.

Sen P., Ciraci S., Buldum A. and Batra I.P. (2001). Structure of aluminum atomic chains, *Phys. Rev. B* **64**, 19, pp. 195420-1–195420-6.

Sheng W-D. and Xia J-B. (1996). A transfer matrix approach to conductance in quantum waveguides, *J. Phys.: Condens. Matter* **8**, pp. 3635–3645.

Shirley E.L., Allan D.C., Martin R.M. and Joannopoulos J.D. (1989). Extended norm-conserving pseudopotentials, *Phys. Rev. B* **40**, 6, pp. 3652–3660.

Sigalas M., Bacalis N.C., Papaconstantopoulos D.A., Mehl M.J. and Switendick A.C. (1990). Total-energy calculations of solid H, Li, Na, K, Rb, and Cs, *Phys, Rev. B* **42**, 18, pp. 11637-11643.

Sim H.-S., Lee H.-W. and Chang K.J. (2001). Even-odd behavior of conductance in monatomic sodium wires, *Phys. Rev. Lett.* **87**, 9, pp. 096803-1–096803-4.

Sim H.-S., Lee H.-W. and Chang K.J. (2002). Even-odd behavior and quantization of conductance in monovalent atomic wire, *Physica E* **14**, pp. 347–354.

Sleijpen G.L.G. and Fokkema D.R. (1993). BiCGSTAB(l) for linear equations involving unsymmetric matrices with complex spectrum, *Electron. Trans. Numer. Anal.*, **1**, pp. 11–32.

Smit R.H.M., Untiedt C., Rubio-Bollinger G., Segers R.C. and van Ruitenbeek J.M. (2003). Observation of a parity oscillation in the conductance of atomic wires, *Phys. Rev. Lett.* **91**, 7, pp. 076805-1–076805-4.

Starkloff Th. and Joannopoulos J.D. (1977). Local pseudopotential theory for transition metals, *Phys. Rev. B* **16**, 12, pp. 5212–5215.

Stearns M.B. (1986), Fe, Co, Ni. In *Magnetic Properties of 3d, 4d and 5d Elements, Alloys and Compounds, Landolt-Börnstein: New Series, Group III, Vol. 19A,* (eds.) Hellwege K.-H. and Madelung O. (Berlin, Springer).

Stechel E.B., Williams A.R. and Feibelman P.J. (1994). *N*-scaling algorithm for density-functional calculations of metals and insulators, *Phys. Rev. B* **49**, 15, pp. 10088–10101.

Stephan U., Drabold D.A. and Martin R.M. (1998). Improved accuracy and acceleration of variational order-N electronic-structure computations by projection techniques, *Phys Rev. B* **58**, 20, pp. 13472–13481.

Stiles M.D. and Hamann D.R. (1988). Ballistic electron transmission through interfaces, *Phys. Rev. B* **38**, 3, pp. 2021–2037.

Sudoh K. and Iwasaki H. (2000). Nanopit formation and manipulation of steps on Si(001) at high temperatures with a scanning tunneling microscope, *Jpn. J. Appl. Phys.* **39**, 7B, pp. 4621–4623.

Sugino O. and Miyamoto Y. (1999). Density-functional approach to electron dynamics: Stable simulation under a self-consistent field, *Phys. Rev. B* **59**, 4, pp. 2579–2586.

Szabo A. and Ostlund N.S. (1989). *Modern Quantum Chemistry*, (New York, McGraw-Hill).

Takagaki Y. and Ferry D.K. (1992). Electronic conductance of a two-dimensional electron gas in the presence of periodic potentials, *Phys. Rev. B* **45**, 15, pp. 8506–8515.

Takahashi H., Hashimoto H. and Nitta T. (2003). Quantum mechanical/molecular mechanical studies of a novel reaction catalyzed by proton transfers in ambient and supercritical states of water, *J. Chem. Phys.* **119**, 15, pp. 7964–7971.

Takahashi H., Hisaoka S. and Nitta T. (2002). Ethanol oxidation reactions catalyzed by water molecules: $CH_3CH_2OH + nH_2O \rightarrow CH_3CHO + H_2 + nH_2O$ (n = 0, 1, 2), *Chem. Phys. Lett.* **363**, 1-2, pp. 80–86.

Takahashi H., Hori T., Hashimoto H. and Nitta T. (2001). A hybrid QM/MM method employing real space grids for QM water in the TIP4P water solvents, *J. Comput. Chem.*, **22**, 12, pp. 1252–1261.

Tamura R. (2003). Resonant spin current in nanotube double junctions, *Phys. Rev. B* **67**, 12, pp. 121408-1–121408-4.

Taraschi G., Mozos J.-L., Wan C.C., Guo H. and Wang J. (1998). Structural and transport properties of aluminum atomic wires, *Phys. Rev. B* **58**, 19, pp. 13138-13145.

Taylor J., Guo H. and Wang J. (2001). *Ab initio* modeling of quantum transport properties of molecular electronic devices, *Phys. Rev. B* **63**, 24, pp. 245407-1–245407-13.

Teter M.P., Payne M.C. and Allan D.C. (1989). Solution of Schrödinger's equation for large systems, *Phys. Rev. B* **40**, 18, pp. 12255–12263.

Thomas L.H. (1927). The calculation of atomic fields, *Proc. Camb. Phil. Soc.* **23**, pp. 542–548.

Thygesen K.S., Bollinger M.V. and Jacobsen K.W. (2003). Conductance calculations with a wavelet basis set, *Phys. Rev. B* **67**, 11, pp. 115404-1–115404-11.

Thygesen K.S. and Jacobsen K.W. (2003). Four-atom period in the conductance of monatomic al wires, *Phys. Rev. Lett.* **91**, 14, pp. 146801-1–146801-4.

Todorov T.N., Briggs G.A.D. and Sutton A.P. (1993). Elastic quantum transport through small structures, *J. Phys.: Condens. Matter* **5**, 15, pp. 2389–2406.

Torres J.A., Tosatti E., Corso A.D., Ercolessi F., Kohanoff J.J., Tolla F.D.D. and Soler J.M. (1999). The puzzling stability of monatomic gold wires, *Surf.*

Sci. **426**, pp. L441-L446.

Tosatti E., Prestipino S., Kostlmeier S., Corso A.D. and Tolla F.D.D. (2001). String tension and stability of magic tip-suspended nanowires, *Science* **291**, pp. 288–290.

Troullier N. and Martins J.L. (1991). Efficient pseudopotentials for plane-wave calculations, *Phys. Rev. B* **43**, 3, pp. 1993–2006.

Tsong T.T. (1990). *Atom Probe Field Ion Microscopy* (Cambridge, Cambridge University Press).

Tsukamoto S., Aono M. and Hirose K. (2002). Sudden suppression of electron-transmission peaks in finite-biased nanowires, *Jpn. J. Appl. Phys.* **41**, 12, pp. 7491-7495.

Tsukamoto S. and Hirose K. (2002). Electron-transport properties of Na nanowires under applied bias voltages, *Phys. Rev. B* **66**, 16, pp. 161402-1–161402-4.

Tsukamoto S., Ono T., Fujimoto Y., Inagaki K., Goto H. and Hirose K. (2001). Geometry and conduction of an infinite single-row gold wire, *Mater. Trans. JIM* **42**, 11, pp. 2257–2260.

Van der Vorst H.A. (1992). Bi-CGSTAB: A fast and smoothly converging variant of Bi-CG for the solution of nonsymmetric linear systems, *SIAM J. Sci. Stat. Comput.* **13**, pp. 631–644.

van Ruitenbeek J.M. (2000). Conductance quantization in metallic point contacts. In *Metal Clusters at Surfaces Structure, Quantum Properties, Physical Chemistry*, (ed.) Meiwes-Broer K.H. (Berlin, Springer).

van Wees B.J., van Houten H., Beenakker C.W.J., Williamson J.G., Kouwenhoven L.P., van der Marel D. and Foxon C.T. (1988). Quantized conductance of point contacts in a two-dimensional electron gas, *Phys. Rev. Lett.* **60**, 9, pp. 848–850.

Vanderbilt D. (1990). Soft self-consistent pseudopotentials in a generalized eigenvalue formalism, *Phys. Rev. B* **41**, 11, pp. 7892–7895.

Vigneron J.P. and Lambin Ph. (1979). A continued-fraction approach for the numerical determination of one-dimensional band structures, *J. Phys. A: Math. Gen.* **12**, 11, pp. 1961–1970.

von Barth U. (1984). An overview of density-functional theory. In *Many Body Phenomena at Surfaces*, (eds.) Langreth D. and Suhl H. (New York, Academic).

Wachutka G. (1986). New layer method for the investigation of the electronic properties of two-dimensional periodic spatial structures: First applications to copper and aluminum, *Phys. Rev. B* **34**, 12, pp. 8512–8527.

Wagner F., Laloyaux T. and Scheffler M. (1998). Errors in Hellmann–Feynman forces due to occupation-number broadening and how they can be corrected, *Phys. Rev. B* **57**, 4, pp. 2102–2107.

Wan C.C., Mozos J.L., Taraschi G., Wang J. and Guo H. (1997). Quantum transport through atomic wires, *Appl. Phys. Lett.* **71**, 3, pp. 419–421.

Wang B. and Stott M.J. (2003). First-principles local pseudopotentials for group-IV elements, *Phys. Rev. B* **68**, 19, pp. 195102-1–195102-6.

Wang B., Wang J. and Guo H. (1999). Current partition: A nonequilibrium green's function approach, *Phys. Rev. Lett.* **82**, 2, pp. 398–401.

Wang J. and Beck T.L. (2000). Efficient real-space solution of the Kohn–Sham equations with multiscale techniques, *J. Chem. Phys.* **112**, 21, pp. 9223–9228.

Wang L.W. and Zunger A. (1994). Large scale electronic structure calculations using the Lanczos method, *Compt. Mater. Sci.* **2**, pp. 326–340.

Wang L.W. and Zunger A. (1995). Local-density-derived semiempirical pseudopotentials, *Phys. Rev. B* **51**, 24, pp. 17398–17416.

Wang Y., Stocks G.M., Shelton W.A., Nicholson D.M.C., Szotek Z. and Temmerman W.M. (1995). Order-N multiple scattering approach to electronic structure calculations, *Phys. Rev. Lett.* **75**, 15, pp. 2867–2870.

Watanabe S., Ichimura M., Onogi T., Ono Y.A., Hashizume T. and Wada Y. (1997). Theoretical study of Ga adsorbates around dangling-bond wires on an H-terminated Si surface: Possibility of atomic-scale ferromagnets, *Jpn. J. Appl. Phys.* **36**, 7B, pp. L929-L932.

Watanabe K. and Satoh T. (1993). Electric field induced surface electronic structures of Si(100) surface, *Surf. Sci.* **287-288**, pp. 502–505.

Watanabe N. and Tsukada M. (2000a). Fast and stable method for simulating quantum electron dynamics, *Phys. Rev. E* **62**, 2, pp. 2914–2923.

Watanabe N. and Tsukada M. (2000b). Finite element approach for simulating quantum electron dynamics in a magnetic field, *J. Phys. Soc. Jpn.* **69**, 9, pp. 2962–2968.

Weinert M. and Davenport J.W. (1992). Fractional occupations and density-functional energies and forces, *Phys. Rev. B* **45**, 23, pp. 13709–13712.

Wentzcovitch R.M., Martins J.L. and Allen P.B. (1992). Energy versus free-energy conservation in first-principles molecular dynamics, *Phys. Rev. B* **45**, 19, pp. 11372–11374.

Wesseling P. (1991). *An Introduction to Multigrid Methods* (New York, Wiley).

Williams A.R., Feibelman P.J. and Lang N.D. (1982). Green's-function methods for electronic-structure calculations, *Phys. Rev. B* **26**, 10, pp. 5433–5444.

Wortmann D., Ishida. H and Blügel S. (2002). Ab initio Green-function formulation of the transfer matrix: Application to complex band structures, *Phys. Rev. B* **65**, 16, pp. 165103-1–165103-10.

Xue Y., Datta S. and Ratner M.A. (2002). First-principles based matrix Green's function approach to molecular electronic devices: general formalism, *Chem. Phys.* **281**, pp. 151–170.

Yang W. (1997). Absolute-energy-minimum principles for linear-scaling electronic-structure calculations, *Phys. Rev. B* **56**, 15, pp. 9294–9297.

Yanson A.I. (2001). Atomic chains and electronic shells: Quantum mechanisms for the formation of nanowires, Ph.D. Thesis, Universiteit Leiden, The Netherlands.

Yanson A.I., Rubio Bollinger G., van den Brom H.E., Agraït N. and van Ruitenbeek J.M. (1998). Formation and manipulation of a metallic wire of single gold atoms, *Nature* **395**, pp. 783–785.

Yin M.T. and Cohen M.L. (1982). Theory of *ab initio* pseudopotential calculations, *Phys. Rev. B* **25**, 12, pp. 7403–7412.

York D.M., Lee T.-S. and Yang W. (1998). Quantum mechanical treatment of biological macromolecules in solution using linear-scaling electronic structure methods, *Phys. Rev. Lett.* **80**, 22, pp. 5011–5014.

Zhu W., Huang Y., Kouri D.J., Arnold M. and Hoffman D.K. (1994). Time-independent wave-packet forms of Schrödinger and Lippmann-Schwinger equations, *Phys. Rev. Lett.* **72**, 9, pp. 1310–1313.

Zunger A. and Cohen M.L. (1979). First-principles nonlocal-pseudopotential approach in the density-functional formalism. II. Application to electronic and structural properties of solids, *Phys. Rev. B* **20**, 10, pp. 4082–4108.

Index

249

RETURN TO: PHYSICS LIBRARY

351 LeConte Hall   510-642-3122

LOAN PERIOD 1 **1-MONTH**	2	3
4	5	6

ALL BOOKS MAY BE RECALLED AFTER 7 DAYS.

Renewable by telephone.

DUE AS STAMPED BELOW.

This book will be held in PHYSICS LIBRARY until JUL 2 5 2005 MAR 0 8 2006		
		•

FORM NO. DD 22
500   4-03

UNIVERSITY OF CALIFORNIA, BERKELEY
Berkeley, California 94720–6000